DATE DUE		
APR - 3 1998			
APR 2 7 1998			
ILL			
842570			
C O F			
8/25/00			
NOV 2 9 2001			
FEB 2 5 2002			
MAR 1 9 2002			
JAN - 2 2002			
DEC 1 0 2003			

D0042040

Physics and Radiobiology
of Nuclear Medicine

Gopal B. Saha

Physics and Radiobiology of Nuclear Medicine

With 76 Illustrations

Springer-Verlag
New York Berlin Heidelberg London Paris
Tokyo Hong Kong Barcelona Budapest

Gopal B. Saha, Ph.D.
Department of Nuclear Medicine
The Cleveland Clinic Foundation
Cleveland, OH 44195
USA

Library of Congress Cataloging-in-Publication Data
Saha, Gopal B.
    Physics and Radiobiology of Nuclear Medicine / Gopal B. Saha.
        p.   cm.
    Includes bibliographical references and index.
    ISBN 0-387-94036-7—ISBN 3-540-94036-7
    1. Medical Physics.   2. Nuclear medicine.   3. Radiobiology.   I. Title.
    [DNLM.   1. Nuclear Medicine.   2. Biophysics.   3. Radiobiology.   WN
440 S131p]
    R895.S25   1993
    616.07'575—dc20
    DNLM/DLC
    for Library of Congress   93-635

Printed on acid-free paper.

Production managed by Hal Henglein, manufacturing supervised by Vincent R. Scelta.
Typeset by Asco Trade Typesetting Ltd., Hong Kong.
Printed and bound by R. R. Donnelley & Sons, Harrisonburg, VA.
Printed in the United States of America.

9 8 7 6 5 4 3 2 1

ISBN 0-387-94036-7 Springer-Verlag New York Berlin Heidelberg
ISBN 3-540-94036-7 Springer-Verlag Berlin Heidelberg New York

To: My mother, Charubala,
    My father, the late Hriday Ranjan,
    My wife, Sipra
    and
    My children, Prantik and Trina

Preface

A basic knowledge of physics, instrumentation, and radiobiology is essential for nuclear physicians and technologists in the practice of nuclear medicine. The nuclear medicine specialty has matured over the past three decades to the extent that there is an increasing need for certification of physicians and technologists to practice nuclear medicine. Each year many medical residents take the American Board of Nuclear Medicine examination and the American Board of Radiology examination with special competency in Nuclear Radiology, and many technologists take the Registry examination in Nuclear Medicine. All these tests include a good portion of physics, instrumentation, and radiobiology in nuclear medicine. It is mandatory that radiology residents pass the physics section of the American Board of Radiology examination.

This book is primarily addressed to this audience. In addition, anyone interested in the basics of physics, instrumentation, and radiobiology in nuclear medicine should find this book useful.

The book contains 14 chapters. Chapters 1 to 6 deal with the basic properties of the atom and nucleus, radionuclidic decay, statistics of counting, production of radionuclides, and interaction of radiation with matter. In Chapters 7 to 11, various instruments such as ionization chambers, gamma well counters, thyroid probes, gamma cameras, and tomographic scanners (single photon emission computed tomography and positron emission tomography) are described. The principles of operation and various parameters affecting the operation of these instruments are included in these chapters. Chapter 12 presents a brief account of radiobiology, highlighting the effects of radiation on humans. Chapter 13 describes the calculation of internal dosimetry. Various regulations related to radiation protection are presented in Chapter 14. Included at the end of the book are several appendices on various constants, a glossary of terms used in the text, and answers to mathematical problems given at the end of each chapter.

The book is concise but comprehensive, with an emphasis on the basic principles of each topic. Sufficient illustrations have been included to help the reader understand the appropriate subject matter. At the end of each chapter, questions related to the specific topic are provided to provoke the reader to assess the sufficiency of knowledge gained. These questions should be very helpful for those taking the certification examinations.

I do not pretend to be infallible in writing a book of such diversified scientific information. Many errors of both commission and omission may have occurred, and I would appreciate having them brought to my attention by interested readers.

Several individuals were very helpful to me during this project. First and foremost, I am ever grateful to Dr. W.J. MacIntyre of our department, whose perusal of the manuscript and numerous suggestions and ideas were absolutely essential for this book. I thank Dr. M.K. Dewanjee of the University of Miami School of Medicine, Miami, Florida, for many helpful suggestions. Thanks are due to Mr. Bruno Sufka of our department, for clarifying some of the issues related to the computer application. Assistance from all members of our department is greatly appreciated.

I am grateful to Dr. R.T. Go, Chairman of our department, whose continued support and understanding have made my work enjoyable.

I express my heartfelt thanks and appreciation to Ms. Rita Buzzelli for typing the manuscript many times over, graciously, conscientiously, and efficiently, and for her dedicated effort which played a monumental role in bringing this book to fruition.

I thank the publisher, Springer-Verlag, for their sincere cooperation throughout the project.

Finally, my wife Sipra's forbearance and encouragement during this endeavor made work a pleasure.

Gopal B. Saha

Contents

CHAPTER 1

Structure of Matter

Matter and Energy

The existence of the universe is explained by two entities: matter and energy. These two entities are interchangeable and exist in different forms to make up all things visible or invisible in the universe. Whereas matter has a definite size, shape, and form, energy has different forms but no size and shape.

Matter is characterized by its quantity, called the *mass*, and is composed of the smallest unit, the atom. In atomic physics, the unit of mass is the atomic mass unit (amu), which is equal to 1.66×10^{-27} kg.

Energy is the capacity to do work and can exist in several forms: kinetic energy (which is due to the motion of matter); potential energy (which is due to the position and configuration of matter); thermal energy (which is due to the motion of atoms or molecules in matter); electrical energy (which is due to the flow of electrons across an electric potential); chemical energy (which is due to chemical reaction); and radiation (energy in motion). Energy can change from one form to another. Of all these forms, radiation is of great importance in nuclear medicine and, therefore, will be discussed in detail.

Mass and energy are interchangeable, and one is created at the expense of the other. This is predicted by the Einstein's mass–energy relationship:

$$E = mc^2 \tag{1.1}$$

where E is energy in ergs; m is the mass in grams, and c is the velocity of light given as 3×10^{10} cm/sec. This relationship states that everything around us can be classified as matter or energy.

Radiation

Radiation is a form of energy in motion through space. It is emitted by one object and absorbed by another. Radiations are of two types:

1. *Particulate radiations:* Examples of these radiations are energetic electrons, protons, neutrons, α particles, and so forth. They have mass and charge,

except neutrons, which are neutral particles. The velocity of their motion depends on their kinetic energy. The particulate radiations originate from radioactive decay, cosmic rays, nuclear reactions, and so forth.

2. *Electromagnetic radiations:* These radiations are a form of energy in motion that does not have mass and charge and can propagate as either waves or discrete packets of energy, called the *photons* or *quanta*. These radiations travel with the velocity of light. Various examples of electromagnetic radiations include radio waves, visible light, heat waves, γ-radiations, and so forth, and they differ from each other in wavelength and hence in energy. It should be pointed out that the sound waves are not electromagnetic radiations.

The energy E of electromagnetic radiations is given by

$$E = h\nu = \frac{hc}{\lambda} \tag{1.2}$$

where h is the Planck constant given as 6.625×10^{-27} erg·s/cycle, ν is the frequency in hertz (Hz), defined as 1 cycle per second, λ is the wavelength in centimeters, and c is the velocity of light in vacuum, which is equal to nearly 3×10^{10} cm/s.

The unit of electromagnetic radiation is given in electron volts (eV), which is defined as the energy acquired by an electron when accelerated through a potential difference of 1 volt. Using 1 eV $= 1.602 \times 10^{-12}$ erg, Eq. (1.2) becomes

$$E \text{ (eV)} = \frac{1.24 \times 10^{-4}}{\lambda} \tag{1.3}$$

where λ is given in centimeters. Table 1.1 lists the different electromagnetic radiations along with their frequencies and wavelengths.

Table 1.1. Characteristics of different electromagnetic radiations.

Type	Energy (eV)	Frequency (Hz)	Wavelength (cm)
Radio, TV	10^{-10}–10^{-6}	10^{4}–10^{8}	10^{2}–10^{6}
Microwave	10^{-6}–10^{-2}	10^{8}–10^{12}	10^{-2}–10^{2}
Infrared	10^{-2}–1	10^{12}–10^{14}	10^{-4}–10^{-2}
Visible	1–2	10^{14}–10^{15}	10^{-5}–10^{-4}
Ultraviolet	2–100	10^{15}–10^{16}	10^{-6}–10^{-5}
x-Rays and γ-rays	100–10^{7}	10^{16}–10^{21}	10^{-11}–10^{-6}

Table 1.2. Characteristics of electrons and nucleons.

Particle	Charge	Mass (amu)*	Mass (kg)	Mass (MeV)[†]
Electron	−1	0.000549	0.9108×10^{-30}	0.511
Proton	+1	1.00728	1.6721×10^{-27}	938.78
Neutron	0	1.00867	1.6744×10^{-27}	939.07

*amu = 1 atomic mass unit = 1.66×10^{-27} kg = 1/12 of the mass of ^{12}C.
[†] 1 atomic mass unit = 931 MeV.

The Atom

The smallest unit in the composition of matter is the atom. The atom is composed of a nucleus at the center and one or more electrons orbiting around the nucleus. The nucleus consists of protons and neutrons, collectively called *nucleons*. The protons are positively charged particles with a mass of 1.00728 amu, and the neutrons are electrically neutral particles with a mass of 1.00867 amu. The electrons are negatively charged particles with a mass of 0.000549 amu. The protons and neutrons are about 1836 times heavier than the electrons. The number of electrons is equal to the number of protons, thus resulting in a neutral atom of an element. The characteristics of these particles are given in Table 1.2. The size of the atom is about 10^{-8} cm (called the Angstrom, Å), whereas the nucleus has the size of 10^{-13} cm (termed the *fermi*, F). The density of the nucleus is of the order of 10^{14} g/cm^3. The electronic arrangement determines the chemical properties of an element, whereas the nuclear structure dictates the stability and radioactive transformation of the atom.

Electronic Structure of the Atom

Several theories have been put forward to describe the electronic structure of the atom, among which the theory of Niels Bohr, proposed in 1913, is the most plausible one and still holds today. The Bohr's atomic theory states that electrons rotate around the nucleus in discrete energy shells that are stationary and arranged in increasing order of energy. These shells are designated as the K shell, L shell, M shell, N shell, and so forth. When an electron jumps from the upper shell to the lower shell, the difference in energy between the two shells appears as electromagnetic radiations or photons. When an electron is raised from the lower shell to the upper shell, the energy difference is absorbed and must be supplied for the process to occur.

The detailed description of the Bohr's atomic structure is provided by the quantum theory in physics. According to this theory, each shell is designated by a quantum number n, called the *principal quantum number*, and denoted by integers, for example, 1 for the K shell, 2 for the L shell, 3 for the M shell, 4 for the N shell and 5 for the O shell. Each energy shell is subdivided into subshells or orbitals, which are designated as s, p, d, f, and so on. For a principal quantum number n, there are n orbitals in a given shell. These orbitals are assigned the *azimuthal quantum numbers*, l, which represent the electron's angular momentum and can assume numerical values of $l = 0, 1, 2 \ldots n - 1$. Thus for the s orbital, $l = 0$; the p orbital, $l = 1$; the d orbital, $l = 2$; the f orbital, $l = 3$; and so forth. According to this description, the K shell has one orbital, designated as $1s$; the L shell has two orbitals, designated as $2s$ and $2p$, and so forth. The orientation of the electron's magnetic moment in a magnetic field is described by the magnetic quantum number, m. The values of m can be $m = -l, -(l - 1), \ldots, 0, \ldots (l - 1), l$. Each electron rotates about its own axis clockwise or anticlockwise, and the spin quantum number, s ($s = -1/2$ or $+1/2$) is assigned to each electron to specify this rotation.

The electron configuration of the atoms of different elements is governed by the following rules:

1. No two electrons can have the same values for all four quantum numbers in a given atom.
2. The orbital of the lowest energy will be filled in first, followed by the next higher energy orbital. The relative energies of the orbitals are $1s < 2s < 2p < 3s < 3p < 4s < 3d < 4p < 5s < 4d < 5p < 6s < 4f < 5d < 6p < 7s$. This order of energy is valid for lighter elements and is somewhat different in heavier elements.
3. There can be a maximum of $2(2l + 1)$ electrons in each orbital.
4. For given values of n and l, each of the available orbitals is first singly occupied such that no electron pairing occurs. Only when all orbitals are singly occupied does electron pairing take place.
5. Each energy shell contains a maximum of $2n^2$ electrons.

The hydrogen atom has one proton in the nucleus and one electron in the orbit. Its electronic structure is represented as $1s^1$. The helium atom has two electrons, which are accommodated in the $1s$ orbital, and thus has the structure of $1s^2$. Now let us consider the structure of $^{16}_{8}O$, which has eight electrons. The first two electrons will fill the 1s orbital. The next two electrons will go to the 2s orbital. There are three p orbitals, designated as p_x, p_y, p_z, which will be occupied by three electrons individually. The eighth electron will occupy the p_x orbital pairing with the electron already in it. Thus, the electronic configuration of $^{16}_{8}O$ is given by $1s^2 \, 2s^2 \, 2p^4$.

The electron configurations in different orbitals and shells are illustrated in Table 1.3, and the structure of $_{28}Ni$ is shown in Figure 1.1.

The electronic structure of the atom characterizes the chemical properties of elements. The outermost shell in the most stable and chemically inert elements such as neon, argon, krypton, and xenon has the electronic structure of ns^2np^6. Helium, although a noble gas, has the $1s^2$ configuration. Elements having electronic configurations different from that of the noble gases

Table 1.3. Electron configurations in different energy shells.

Principal shell	Principal quantum number (n)	Orbital (l)	No. of electrons = 2(2l + 1) in each orbital	$2n^2$
K	1	s(0)	2	2
L	2	s(0)	2	
		p(1)	6	8
M	3	s(0)	2	
		p(1)	6	
		d(2)	10	18
N	4	s(0)	2	
		p(1)	6	
		d(2)	10	
		f(3)	14	32
O	5	s(0)	2	
		p(1)	6	
		d(2)	10	
		f(3)	14	
		g(4)	18	50

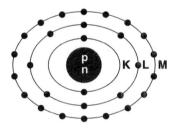

Fig. 1.1. The electronic configuration of $_{28}$Ni. The K shell has 2 electrons, the L shell has 8 electrons, and the M shell has 18 electrons.

either lose or gain electrons to achieve the structure ns^2np^6 of the nearest noble gas atom. The electrons in these shells are called the *valence electrons* and are primarily responsible for the chemical bond formation.

Electrons in different shells are bound to the nucleus by binding energy. The binding energy of an electron is defined as the energy that is required to be supplied to remove it completely from a shell. The binding energy of the electron is the greatest in the K shell and decreases with higher shells such as L, M, and so on. The binding energy also increases with increasing atomic number of the elements. Thus, the K-shell binding energy (21.05 keV) of technetium, with atomic number 43, is higher than the K-shell binding energy (1.08 keV) of sodium, with atomic number 11. The K-shell binding energy of electrons in several elements are: carbon, 0.28 keV; gallium, 10.37 keV; technetium, 21.05 keV; indium, 27.93 keV; iodine, 33.16 keV; lead, 88.00 keV.

When an electron is removed completely from an atom, the process is called *ionization*. The atom is said to be ionized and becomes an ion. On the other hand, when the electron is raised from a lower energy shell to an upper energy shell, the process is called *excitation*. Both ionization and excitation processes require a supply of energy from outside the atom such as heating, applying an electric field, and so forth. In the excited atoms, electrons jump from the upper energy shell to the lower energy shell to achieve stability. The difference in energy appears as electromagnetic radiations or photons. Thus, if the binding energy of K-shell electrons in, say, bromine is 13.5 keV and the L-shell binding energy is 1.8 keV, the transition of electrons from the L shell to the K shell will occur with the emission of 11.7 keV (13.5 − 1.8 = 11.7 keV) photons. As we shall see later, these radiations are called the *characteristic x-rays* of the product atom.

Structure of the Nucleus

As already stated, the nucleus of an atom is composed of protons and neutrons. The number of protons is called the *atomic number* of the element and denoted by Z. The number of neutrons is denoted by N, and the sum of the protons and neutrons, $Z + N$, is called the *mass number*, denoted by A. The symbolic representation of an element, X, is given by $^A_Z X_N$. For example, sodium has 11 protons and 12 neutrons with a total of 23 nucleons. Thus, it is represented as $^{23}_{11}\mathrm{Na}_{12}$. However, the atomic number Z of an element is known, and N can be calculated as $A − Z$; therefore, it suffices to simply write $^{23}\mathrm{Na}$ (or Na-23).

To explain the various physical observations related to the nucleus of an atom, two models for the nuclear structure have been proposed: the liquid drop model and the shell model. The liquid drop model was introduced by Niels Bohr, which assumes a spherical nucleus composed of closely packed nucleons. This model explains various phenomena, such as nuclear density, energetics of particle emission in nuclear reactions, and fission of heavy nuclei.

In the shell model, both protons and neutrons are arranged in discrete energy shells in a manner similar to the electron shells of the atom in the Bohr atomic theory. Similar to the electronic configuration of the noble gas atoms, nuclei with 2, 8, 20, 28, 50, 82, or 126 protons or neutrons are found to be very stable. These nucleon numbers are called the *magic numbers*.

It is observed that atomic nuclei containing an odd number of protons or neutrons are normally less stable than those with an even number of protons or neutrons. Thus, nuclei with even numbers of protons and neutrons are more stable, whereas those with odd numbers of protons and neutrons are less stable. For example, $^{12}\mathrm{C}$ with six protons and six neutrons is more stable than $^{13}\mathrm{C}$ containing six protons and seven neutrons.

There are about 270 stable atoms of naturally occurring elements. The stability of these elements is dictated by the configuration of protons and neutrons. The ratio of the number of neutrons to the number of protons

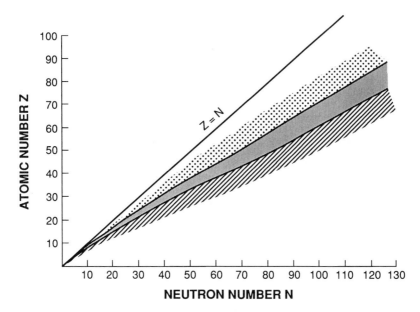

Fig. 1.2. The plot of atomic number (Z) versus the number of neutrons (N) for all nuclides. The proton-rich nuclides fall on the left (dotted) and the neutron-rich nuclides fall on the right (cross-hatched) of the line of stability, indicated by the dark-shaded area. The solid line represents nuclides with $Z = N$.

(N/Z) is an approximate indicator of the stability of a nucleus. The N/Z ratio is 1 in low-Z elements such as $^{12}_{6}C$, $^{14}_{7}N$, and $^{16}_{8}O$, but it increases with increasing atomic number of elements. For example, it is 1.40 for $^{127}_{53}I$ and 1.54 for $^{208}_{82}Pb$. The plot of the atomic number versus the neutron number of all nuclides is shown in Figure 1.2. All stable nuclear species fall on or around what is called the *line of stability*. The nuclear species on the left side of the line have fewer neutrons and more protons; that is, they are proton-rich. On the other hand, those on the right side of the line have fewer protons and more neutrons; that is, they are neutron-rich. The nuclides away from the line of stability are unstable and disintegrate to achieve stability.

Nuclear Binding Energy

According to the classical electrostatic theory, the nucleus of an atom cannot exist as a single entity, because of the electrostatic repulsive force among the protons in the nucleus. The stability of the nucleus is explained by the existence of a strong binding force called the *nuclear force*, which overcomes the repulsive force of the protons. The nuclear force is effective equally among all nucleons and exists only in the nucleus, having no influence outside the nucleus. The short range of the nuclear force leads to a very small size ($\sim 10^{-13}$ cm) and very high density ($\sim 10^{14}$ g/cm^3) of the nucleus.

The mass M of a nucleus is always less than the combined masses of the nucleons A in the nucleus. The difference in mass ($M - A$) is termed the *mass defect*, which has been used as binding energy for all nucleons in the nucleus. The average binding energy of a nucleon is equal to the total binding energy (calculated from the mass defect) divided by the number of nucleons. It is of the order of 6–9 MeV, although the binding energy of an individual nucleon has a definite value, depending on the shell it occupies. The binding energy of a nucleon must be supplied to completely remove it from the nucleus. Note that whereas the binding energy of the nucleons is in the megaelectron volt (MeV) range, the electron binding energy in the atomic orbital is of the order of kiloelectron volts (keV), a factor of 1000 lower.

Nuclear Nomenclature

A *nuclide* is an atomic species with a definite number of protons and neutrons arranged in a definite order in the nucleus.

Radionuclides are those nuclides that are unstable and thus decay by emission of particles or electromagnetic radiations or by spontaneous fission.

Isotopes are the nuclides having the same atomic number Z but different mass number A. Isotopes exhibit the same chemical properties. Examples of carbon isotopes are $^{11}_{6}C$, $^{12}_{6}C$ and $^{13}_{6}C$.

Isotones are the nuclides having the same number of neutrons N but different numbers of protons. Examples of isotones are: $^{134}_{55}Cs$, $^{133}_{54}Xe$ and $^{132}_{53}I$, each having 79 neutrons.

Isobars are the nuclides with the same number of nucleons: that is, the same mass number A, but a different combination of protons and neutrons. For example: ^{82}Y, ^{82}Sr, ^{82}Rb, and ^{82}Kr are all isobars having the mass number 82.

Isomers are the nuclides with the same number of protons and neutrons, but having different energy states and spins. ^{99}Tc and ^{99m}Tc are isomers of the same nuclide. Individual nuclides can exist in different energy states above the ground state due to excitation. These excited states are called the *isomeric states*, which can have a lifetime varying from picoseconds to years. When the isomeric states are long-lived, they are referred to as *metastable states*. These states are denoted by "m" as in ^{99m}Tc.

Chart of the Nuclides

Nearly 3000 nuclides, both stable and unstable, are arranged in the form of a chart, called the *chart of the nuclides*, a section of which is presented in Figure 1.3. Each square in the chart represents a specific nuclide, containing various information such as the half-life, type and energy of radiations, and so forth of the nuclide, and neutron capture cross section of the stable nuclide. The

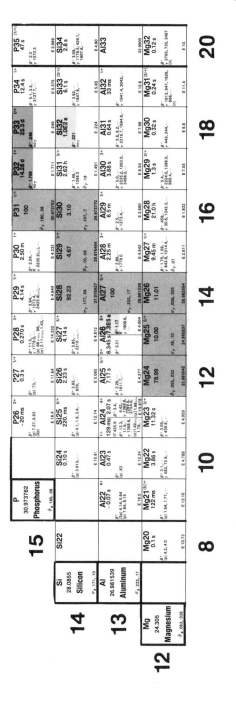

Fig. 1.3. A section of the chart of nuclides. (Courtesy of Knolls Atomic Power Laboratory, Schenectady, New York, operated by the General Electric Company for Naval Reactors, the U.S. Department of Energy.)

nuclides are arranged in increasing neutron number N horizontally and in increasing proton number Z vertically. Each horizontal group of squares contains all isotopes of the same element, whereas the vertical group contains all isotones with the same number of neutrons. For isomers, the square is subdivided into sections representing each isomer.

Questions

1. If a mass of matter (m) is converted to electromagnetic radiation, what should be the energy of this radiation?
2. Describe the Bohr's atomic theory in terms of the electronic configuration of the atom.
3. What is the difference between the orbital electron binding energy and the nuclear binding energy of an atom?
4. Define the mass defect and mass number of an atom. What does the mass defect account for?
5. Write the electronic configuration of ^{99m}Tc and ^{131}I.
6. How many electrons can the $3d$ orbital contain?
7. The electron binding energy of the K shell in an atom is higher than that of the L shell. True or False?
8. What is the difference between ionization and excitation of an atom?
9. What is a metastable state of a nuclide? How is it designated?

Suggested Readings

Friedlander G, Kennedy TW, Miller JM. *Nuclear and Radiochemistry.* 3rd ed. New York: Wiley; 1981.

Harvey BG. *Introduction to Nuclear Physics and Chemistry.* 2nd ed. Englewood Cliffs, NJ: Prentice-Hall; 1969.

Sorensen JA, Phelps ME. *Physics in Nuclear Medicine.* 2nd ed. New York: Grune & Stratton; 1987.

Sprawls P, Jr. *Physical Principles of Medical Imaging.* Rockville, Md: Aspen Publishers, Inc; 1987.

CHAPTER 2

Radioactive Decay

In 1896, Henri Becquerel first discovered natural radioactivity in potassium uranyl sulfate. Artificial radioactivity was not produced until 1934, when I. Curie and F. Joliot made boron, aluminum, and magnesium radioactive by bombarding them with α particles from polonium. This introduction of artificial radioactivity prompted the invention of cyclotrons and reactors in which many radionuclides are now produced. So far, more than 2700 radionuclides have been artificially produced and characterized in terms of their physical properties.

Radionuclides are unstable and decay by emission of particle or γ-radiation to achieve stable configuration of protons and neutrons in the nucleus. As already mentioned, the stability of a nuclide is determined by the N/Z ratio of the nucleus. Thus, as will be seen later, whether a nuclide will decay by a particular particle emission or γ-ray emission is determined by the N/Z and/or excitation energy of the nucleus. Radionuclides can decay by one or more of the five modes: alpha (α) decay, beta (β^-) decay, positron (β^+) decay, electron capture (EC) decay, and isomeric transition (IT). In all decay modes, energy and mass are conserved. Different decay modes of radionuclides are described later in detail.

Isomeric Transition

As previously mentioned, a nucleus can exist in different energy or excited states above the ground state, which is considered as the state involving the arrangement of protons and neutrons with the least amount of energy. These excited states are called the *isomeric states* and have lifetimes of fractions of picoseconds to many years. When isomeric states are long-lived, they are referred to as *metastable states* and denoted by "m" as in ^{99m}Tc. A nucleus in an excited state gives off its energy and returns to the ground state in several ways. The most common mode of a nuclear transition from an upper excited state to a lower excited state is by the emission of an electromagnetic radiation, called the gamma (γ)-radiation. Such transitions are called *iso-*

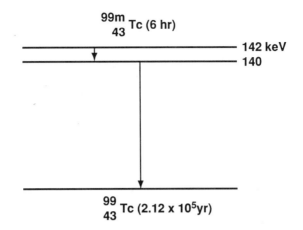

Fig. 2.1. Isomeric transition of ^{99m}Tc. Ten percent of the decay follows internal conversion.

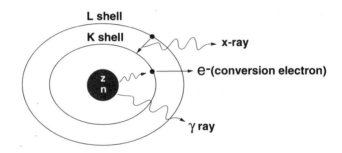

Fig. 2.2. Internal conversion process. The excitation energy of the nucleus is transferred to a K-shell electron, which is then ejected, and the K-shell vacancy is filled by an electron from the L shell. The energy difference between the L shell and K shell appears as the characteristic K x-ray.

meric transitions (IT) and may occur in several steps involving intermediate excited states instead of directly to the ground state. A typical isomeric transition of ^{99m}Tc is illustrated in Figure 2.1.

As will be seen later, a parent radionuclide may reach an isomeric state of the product nucleus after α-particle or β-particle decay, in which case, the isomeric state returns to the ground state by γ-ray emission.

An alternative to the isomeric transition is the *internal conversion* process. The excited nucleus transfers the excitation energy to an orbital electron— preferably the K-shell electron—of its own atom, which is then ejected from the shell, provided the excitation energy is greater than the binding energy of the electron (Fig. 2.2). The ejected electron is called the *conversion electron* and carries the kinetic energy equal to $E_\gamma - E_B$, where E_γ is the excitation energy and E_B is the binding energy of the electron. Even though the K-shell electrons are more likely to be ejected because of the proximity to the nucleus, the electrons from the L shell, M shell, and so forth also may undergo the internal conversion process. The ratio of the number of conversion electrons (N_e) to the number of observed γ-radiations (N_γ) is referred to as the *conversion coefficient*, given as $\alpha = N_e/N_\gamma$. The conversion coefficients are subscripted as α_K, α_L, α_M ... depending on which shell the electron is ejected from. The total conversion coefficient α_T is then given by

$$\alpha_T = \alpha_K + \alpha_L + \alpha_M + \cdots$$

Problem 2.1

If the total conversion coefficient (α_T) is 0.11 for the 140-keV γ-rays of ^{99m}Tc, calculate the percentage of 140-keV γ-radiations available for imaging.

Answer

$$\alpha_T = \frac{N_e}{N_\gamma} = 0.11$$

$$N_e = 0.11\ N_\gamma$$

Total number of disintegrations

$$= N_e + N_\gamma$$
$$= 0.11\ N_\gamma + N_\gamma$$
$$= 1.11\ N_\gamma$$

Thus, the percentage of γ-radiations

$$= \frac{N_\gamma}{1.11\ N_\gamma} \times 100$$

$$= \frac{1}{1.11} \times 100$$

$$= 90\%$$

An internal conversion process leaves an atom with a vacancy in one of its shells, which is filled most frequently by an electron from the next higher shell. Such situations may also occur in nuclides decaying by electron capture (see later). When an L electron fills in a K-shell vacancy, the energy difference between the K shell and the L shell appears as a characteristic K x-ray. Alternatively, this transition energy may be transferred to an orbital electron, which is emitted with a kinetic energy equal to the characteristic x-ray energy minus its binding energy. These electrons are called *Auger electrons*, and the process is termed the *Auger process*, analogous to internal conversion. Because the characteristic x-ray energy (energy difference between the two shells) is always less than the binding energy of the K-shell electron, the latter cannot undergo the Auger process and cannot be emitted as an Auger electron.

The vacancy in the shell resulting from an Auger process is filled by the transition of an electron from the next upper shell, followed by emission of similar characteristic x-rays and/or Auger electrons. The fraction of vacancies in a given shell that are filled with accompanying characteristic x-ray emissions is called the *fluorescence yield*, and the fractions that are filled by the Auger processes is the *Auger yield*. The fluorescence yield increases with the increasing atomic number of the atom.

Alpha (α) Decay

The α-decay occurs mostly in heavy nuclides such as uranium, radon, plutonium, and so forth. Beryllium-8 is the lightest nuclide that decays by α particle emission. The α particles are basically helium ions with two protons and two neutrons in the nucleus and two electrons removed from the helium atom. After α decay, the atomic number of the nucleus is reduced by 2 and the mass number by 4.

$$^{222}_{86}\text{Rn} \rightarrow {}^{218}_{84}\text{Po} + \alpha$$

The α particles from a given radionuclide all have discrete energies corresponding to the decay of the initial nuclide to a particular energy level of the product (including, of course, its ground state). The energy of the α particles is, as a rule, equal to the energy difference between the two levels and ranges from 1 to 10 MeV. The high-energy α particles normally originate from the short-lived radionuclides and vice versa. The range of the α particles is very short in matter and is approximately 0.03 mm in body tissue. The α particles can be stopped by a piece of paper, a few centimeters of air, and gloves.

Beta (β^-) Decay

When a radionuclide is neutron rich—that is, the N/Z ratio is greater than that of the nearest stable nuclide—it decays by the emission of a β^- particle (note that it is an electron*) and an antineutrino, $\bar{\nu}$. In the β^--decay process, a neutron is converted to a proton, thus raising the atomic number Z of the product by 1. Thus:

$$n \rightarrow p + \beta^- + \bar{\nu}$$

The difference in energy between the parent and daughter nuclides is called the *transition or decay energy*, denoted by E_{max}. The β^- particles carry E_{max}

---

*The difference between a β particle and an electron is that a β^- particle originates from the nucleus, and an electron originates from the extranuclear electron orbitals.

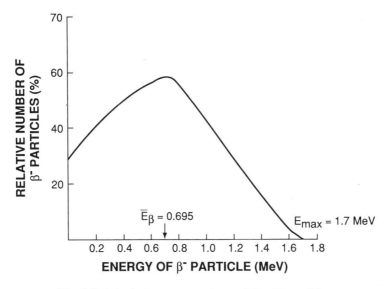

Fig. 2.3. A typical energy spectrum of the β^- particles.

or part of it, exhibiting a spectrum of energy as shown in Figure 2.3. The average energy of the β^- particles is about one-third of E_{max}. This observation indicates that β^- particles often carry only a part of the transition energy, and energy is not conserved in β^- decay. To satisfy the law of energy conservation, a particle called the *antineutrino*, $\bar{\nu}$, with no charge and a negligible mass has been postulated, which carries the remainder of E_{max} in each β^- decay. The existence of antineutrinos has been proven experimentally.

After β^- decay, the daughter nuclide may exist in an excited state, in which case, one or more isomeric transitions or γ-ray emissions will occur to dispose of the excitation energy. In other words, β^- decay is followed by γ-ray emission, if energetically permitted.

The decay process of a radionuclide is normally represented by what is called the *decay scheme*. Typical decay schemes of ^{131}I and ^{99}Mo are shown in Figures 2.4 and 2.5, respectively. The β^- decay is shown by a left-to-right arrow from the parent nuclide to the daughter nuclide, whereas the isomeric transition is displayed by a vertical arrow between the two states. (Note: The β^+ decay is shown by a two-step right-to-left arrow between the two states, the electron capture decay by a right-to-left arrow, and the α-decay by a down arrow). Although it is often said that, ^{131}I emits 364-keV γ-rays, it should be understood that the 364-keV γ-ray belongs to ^{131}Xe as an isomeric state. This is true for all β^-, β^+, or electron capture decays that are followed by γ-ray emission.

Some examples of β^- decay follow:

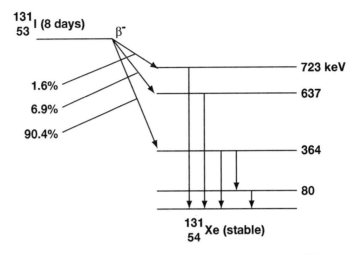

Fig. 2.4. Decay scheme of ^{131}I. Eighty-one percent of the total ^{131}I radionuclides decay by 364-keV γ-ray emission. The 8.0-day half-life of ^{131}I is shown in parenthesis.

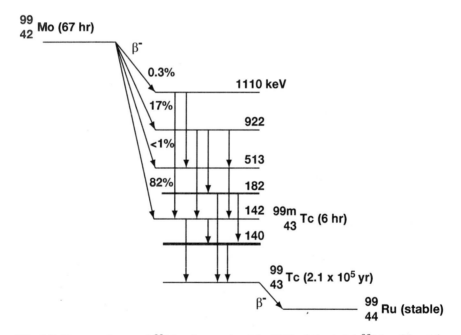

Fig. 2.5. Decay scheme of ^{99}Mo. Approximately 87% of the total ^{99}Mo ultimately decays to ^{99m}Tc, and the remaining 13% decays to ^{99}Tc. A 2-keV transition occurs from the 142-keV level to the 140-keV level. All the 2-keV γ-rays are internally converted. (The energy levels are not shown in scale.)

$$^{99}_{42}\text{Mo} \rightarrow {}^{99\text{m}}_{43}\text{Tc} + \beta^- + \bar{\nu}$$

$$^{131}_{53}\text{I} \rightarrow {}^{131}_{54}\text{Xe} + \beta^- + \bar{\nu}$$

$$^{67}_{29}\text{Cu} \rightarrow {}^{67}_{30}\text{Zn} + \beta^- + \bar{\nu}$$

$$^{90}_{38}\text{Sr} \rightarrow {}^{90}_{39}\text{Y} + \beta^- + \bar{\nu}$$

It should be noted that in β^- decay, the atomic number of the daughter nuclide is increased by 1 and the mass number remains the same.

Positron (β^+) Decay

When a radionuclide is proton rich—that is, the N/Z ratio is low relative to that of the nearest stable nuclide—it can decay by positron (β^+) emission accompanied by the emission of a neutrino (ν), which is an opposite entity of the antineutrino. Positron emission takes place when the energy difference (transition energy) between the parent and daughter nuclides is greater than 1.02 MeV. In β^+ decay, essentially a proton is converted to a neutron plus a positron, thus, decreasing the atomic number Z of the daughter nuclide by 1. Thus,

$$p \rightarrow n + \beta^+ + \nu$$

The requirement of 1.02 MeV for β^+ decay arises from the fact that one electron mass has to be added to a proton to produce a neutron and one

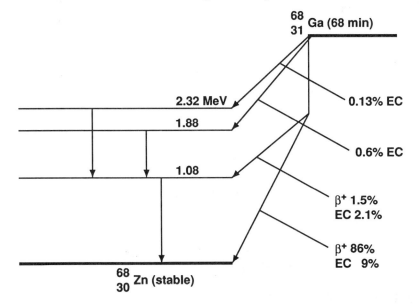

Fig. 2.6. Decay scheme of ^{68}Ga. The positrons are annihilated in medium to give rise to two 511-keV γ-rays emitted in opposite directions.

positron is created. Since each electron or positron mass is equal to 0.511 MeV, one electron and one positron are equal to 1.02 MeV, which is required as a minimum for β^+ decay.

Some examples of β^+ decay follow:

$$^{18}_{9}F \rightarrow {}^{18}_{8}O + \beta^+ + \nu$$

$$^{68}_{31}Ga \rightarrow {}^{68}_{30}Zn + \beta^+ + \nu$$

$$^{13}_{7}N \rightarrow {}^{13}_{6}C + \beta^+ + \nu$$

$$^{15}_{8}O \rightarrow {}^{15}_{7}N + \beta^+ + \nu$$

The energetic β^+ particle loses energy while passing through matter. The range of positrons is short in matter. When it loses almost all of its energy, it combines with an atomic electron of the medium and is annihilated, giving rise to two photons of 511 keV emitted at 180°. These photons are called *annihilation radiations*.

The decay scheme of ^{68}Ga is presented in Figure 2.6. Note that the β^+ decay is represented by a two-step right-to-left arrow.

Electron Capture

Decay by EC is an alternative to the β^+ decay for proton-rich radionuclides with N/Z lower than that of the stable nuclide. In EC decay, an electron from an extranuclear shell, particularly the K shell because of its proximity, is

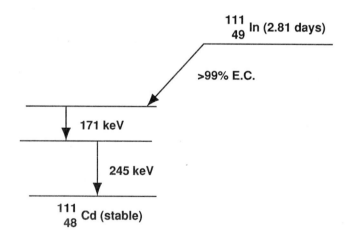

Fig. 2.7. Decay scheme of ^{111}In illustrating the electron capture process. The abundances of 171-keV and 245-keV γ-rays are 90% and 94%, respectively.

captured by a proton in the nucleus, forming a neutron and a neutrino. Thus,

$$p + e^- \rightarrow n + v$$

In this process, the atomic number of the daughter nuclide is lowered by 1. The EC process occurs usually in nuclides having excitation energy less than 1.02 MeV. In nuclides having excitation energy greater than 1.02 MeV, both EC and β^+ decay can occur, although the probability of β^+ decay increases with higher excitation energy. The decay scheme of ^{111}In is shown in Figure 2.7. The EC decay is indicated by a right-to-left arrow. Some examples of EC decay follow:

$$^{111}_{49}\text{In} + e^- \rightarrow {}^{111}_{48}\text{Cd} + v$$

$$^{67}_{31}\text{Ga} + e^- \rightarrow {}^{67}_{30}\text{Zn} + v$$

$$^{125}_{53}\text{I} + e^- \rightarrow {}^{125}_{52}\text{Te} + v$$

$$^{57}_{27}\text{Co} + e^- \rightarrow {}^{57}_{26}\text{Fe} + v$$

$$^{123}_{53}\text{I} + e^- \rightarrow {}^{123}_{52}\text{Te} + v$$

In EC decay, analogous to the situation in internal conversion, a vacancy is created in the shell from which the electron is captured. It is filled in by the transition of an electron from the next upper shell, in which case the difference in energy between the two shells appears as a characteristic x-ray of the daughter nuclide. Also, instead of characteristic x-ray emission, the Auger process can occur, whereby an Auger electron is emitted.

Questions

1. What are the primary criteria for β^+ and β^- decay?
2. If the energy difference between the proton-rich parent nuclide and the daughter nuclide is 1.2 MeV, could the parent radionuclide decay by β^+ decay and/or electron capture? If the energy difference is 0.8 MeV, what should be the mode of decay?
3. If the total conversion coefficient (α_T) of 195-keV γ-rays of a radionuclide is 0.23, calculate the percentage of 195-keV photons available for imaging.
4. Can a K-shell electron be emitted as an Auger electron? Explain.
5. Explain how characteristic x-rays and Auger electrons are emitted.
6. Why is an antineutrino emitted in β^- decay?
7. A K-shell electron is ejected by the internal conversion of a 155-keV γ-ray photon. If the binding energy of the K-shell electron is 25 keV, what is the kinetic energy of the electron?
8. What is the average energy of the β^- particles emitted from a radionuclide?
9. Explain the production of annihilation radiations.

Suggested Readings

Friedlander G, Kennedy JW, Miller JM. *Nuclear and Radiochemistry*. 3rd ed. New York: Wiley; 1981.

Harvey BG. *Introduction to Nuclear Physics and Chemistry*. 2nd ed. Englewood Cliffs, NJ: Prentice-Hall; 1969.

Sorensen JA, Phelps ME. *Physics in Nuclear Medicine*. 2nd ed. New York: Grune & Stratton; 1987.

Sprawls P Jr. *Physical Principles of Medical Imaging*. Rockville, Md: Aspen Publishers, Inc; 1987.

CHAPTER 3

Kinetics of Radioactive Decay

Radioactive Decay Equations

General Equation

As mentioned in Chapter 2, radionuclides decay by α-, β^-- and β^+-particle emission, electron capture, or isomeric transition. The radioactive decay is a random process, and it is not possible to tell which atom from a group of atoms disintegrates at a specific time. Thus, one can only talk about the average number of radionuclides disintegrating during a period of time. This gives the disintegration rate of a particular radionuclide.

The disintegration rate of a radionuclide, that is, the number of disintegrations per unit time, given as $-dN/dt$, is proportional to the total number of radionuclides present at that time. Mathematically,

$$-dN/dt = \lambda N \tag{3.1}$$

where N is the number of radionuclides present, and λ is referred to as the *decay constant* of the radionuclide. As can be seen from Eq. (3.1), it is a small fraction of the radionuclides that decays in a very short period of time. The unit of λ is $(\text{time})^{-1}$. Thus, if λ is 0.2 $\sec^{-1}$ for a radionuclide, then 20% of the radionuclides present will disappear per second.

The disintegration rate $-dN/dt$ is denoted by D. It is also referred to as the *radioactivity* or simply the *activity* and denoted by A. It should be understood from Equation (3.1) that the same amount of radioactivity means the same disintegration rate for any radionuclide, but the total number of atoms present and the decay constants differ for different radionuclides. For example, a radioactive sample A containing 10^6 atoms and with $\lambda = 0.01$ min^{-1} would give the same disintegration rate (10,000 disintegrations per minute) as that by a radioactive sample B containing 2×10^6 atoms and with a decay constant 0.005 min^{-1}.

Now from the preceding discussion, the following equations can be written:

$$D = \lambda N \tag{3.2}$$

$$A = \lambda N \qquad\qquad (3.3)$$

From a knowledge of the decay constant and radioactivity of a radionuclide, one can calculate the total number of atoms or mass of the radionuclides present (using Avogadro's number 1 g·atom $= 6.02 = 10^{23}$ atoms).

Because Eq. (3.1) is a first-order differential equation, the solution of this equation by integration leads to

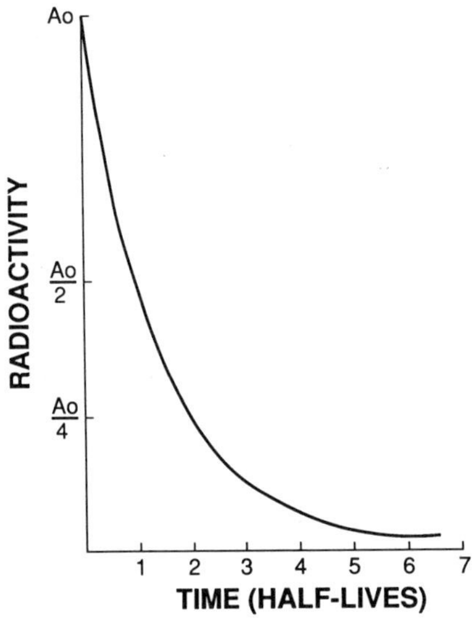

Fig. 3.1. Plot of radioactivity versus time on a linear graph indicting an exponential curve.

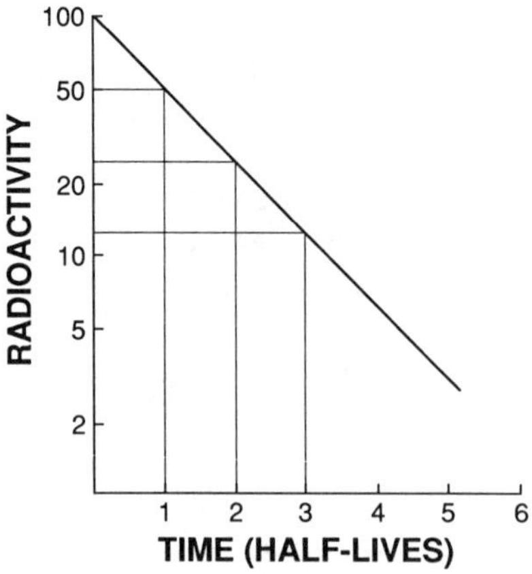

Fig. 3.2. Plot of radioactivity against time on a semilogarithmic graph indicating a straight line. The half-life of the radionuclide can be determined from the slope of the line, which is given as the decay constant λ. Alternatively, an activity and half its value and their corresponding times are read from the plot. The difference in the two time readings gives the half-life.

$$N_t = N_0 e^{-\lambda t} \tag{3.4}$$

where N_0 and N_t are the number of radionuclides at $t = 0$ and time t, respectively. Equation (3.4) is an exponential equation indicating that the radioactivity decays exponentially. By multiplying both sides of Eq. (3.4) by λ, one obtains

$$D_t = D_0 e^{-\lambda t} \tag{3.5}$$

$$A_t = A_0 e^{-\lambda t} \tag{3.6}$$

The factor $e^{-\lambda t}$ is called the *decay factor*. The decay factor becomes $e^{+\lambda t}$ if the activity at time t before $t = 0$ is to be determined. The plot of activity versus time on a linear graph gives an exponential curve, as shown in Figure 3.1. However, if the activity is plotted against time on semilogarithmic paper, a straight line results, as shown in Figure 3.2.

Half-Life

Every radionuclide is characterized by a *half-life*, which is defined as the time required to reduce its initial disintegration rate or activity to one half. It is usually denoted by $t_{1/2}$ and is unique for a radionuclide. It is related to the decay constant λ of a radionuclide by

$$\lambda = \frac{0.693}{t_{1/2}} \tag{3.7}$$

From the definition of half-life, it is understood that A_0 is reduced to $A_0/2$ in one half-life; to $A_0/4$, that is, to $A_0/2^2$ in two half-lives; to $A_0/8$, that is, $A_0/2^3$ in three half-lives; and so forth. In n half-lives of decay, it is reduced to $A_0/2^n$. Thus, the radioactivity A_t at time t can be calculated from the initial radioactivity A_0 by

$$A_t = \frac{A_0}{2^n} = \frac{A_0}{2^{(t/t_{1/2})}} = A_0 \, (0.5)^{t/t_{1/2}} \tag{3.8}$$

where t is the time of decay. Here, $t/t_{1/2}$ can be an integer or a fraction depending on t and $t_{1/2}$. For example, a radioactive sample with $t_{1/2} = 3.2$ days decaying at a rate of 10,000 disintegrations per minute would give, after 7 days of decay, $10{,}000/2^{(7/3.2)} = 10{,}000/2^{2.2} = 10{,}000/4.59 = 2178$ disintegrations per minute.

It should be noted that ten half-lives of decay reduce the radioactivity by a factor of about 1000 ($2^{10} = 1024$), or to 0.1% of the initial activity.

The half-life of a radionuclide is determined by measuring the radioactivity or disintegration rates at different time intervals and plotting them on semilogarithmic paper, as shown in Figure 3.2. An initial activity and half its value are read from the line, and the corresponding times are noted. The difference in time between the two readings gives the half-life of the radionuclide. For a very long-lived radionuclide, the half-life is determined by Eqs. (3.2) or (3.3) from a knowledge of its activity or disintegration rate and the number of atoms present. The number of atoms N can be calculated from the weight W of the radionuclide with atomic weight A and Avogadro's number 6.02×10^{23} atoms per g·atom as follows:

$$N = \frac{W}{A} \times 6.02 \times 10^{23} \tag{3.9}$$

When two or more radionuclides are present in a sample, the measured count of such a sample comprises counts of all individual radionuclides. A semilogarithmic plot of the activity of a two-component sample versus time is shown in Figure 3.3. The half-life of each of the two radionuclides can be determined by what is called the *peeling or stripping method*. In this method, first, the tail part (second component) of the curve is extrapolated as a straight line up to the ordinate, and its half-life can be determined as mentioned previously (e.g., 27 hr). Second, the activity values on this line are subtracted from those on the composite line to obtain the activity values for

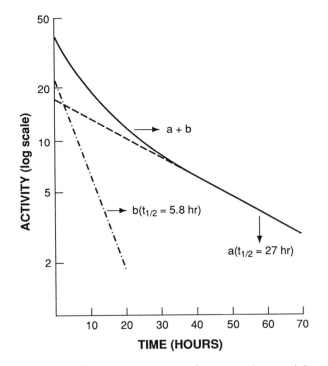

Fig. 3.3. A composite radioactive decay curve for a sample containing two radio-nuclides of different half-lives. The long-lived component (a) has a half-life of 27 hr and the short-lived component (b) has a half-life of 5.8 hr.

the first component. A straight line is drawn through these points, and the half-life of the first component is determined (e.g., 5.8 hr). The stripping method can be applied to more than two components in the similar manner.

Mean Life

Another relevant quantity of a radionuclide is its *mean life*, which is the average lifetime of a group of radionuclides. It is denoted by τ and is related to the decay constant λ and half-life $t_{1/2}$ as follows:

$$\tau = 1/\lambda \tag{3.10}$$

$$\tau = t_{1/2}/0.693 = 1.44\, t_{1/2} \tag{3.11}$$

In one mean life, the activity of a radionuclide is reduced to 37% of its initial value.

Effective Half-Life

As already mentioned, a radionuclide decays exponentially with a definite half-life, which is called the *physical half-life*, denoted by T_p (or $t_{1/2}$). The physical half-life of a radionuclide is independent of its physicochemical conditions. Analogous to physical decay, radiopharmaceuticals administered to humans disappear exponentially from the biologic system through fecal excretion, urinary excretion, perspiration, or other routes. Thus, every radiopharmaceutical has a *biologic half-life* (T_b), which is defined as the time needed for half of the radiopharmaceutical to disappear from the biologic system. It is related to decay constant λ_b by $\lambda_b = 0.693/T_b$.

Obviously, in any biologic system, the loss of a radiopharmaceutical is due to both the physical decay of the radionuclide and the biologic elimination of the radiopharmaceutical. The net or effective rate (λ_e) of loss of radioactivity is then related to λ_p and λ_b by

$$\lambda_e = \lambda_p + \lambda_b \tag{3.12}$$

Because $\lambda = 0.693/t_{1/2}$, it follows that

$$\frac{1}{T_e} = \frac{1}{T_p} + \frac{1}{T_b} \tag{3.13}$$

or,

$$T_e = \frac{T_p \times T_b}{T_p + T_b} \tag{3.14}$$

The effective half-life, T_e, is always less than the shorter of T_p or T_b. For a very long T_p and a short T_b, T_e is almost equal to T_b. Similarly, for a very long T_b and short T_p, T_e is almost equal to T_p.

Units of Radioactivity

The unit of radioactivity is a curie. It was historically defined as the disintegration rate of 1 g of radium, which was measured to be 3.7×10^{10} disintegrations per second. Later, the disintegration rate of 1 g of radium was found to be slightly different from this value, but the original definition of curie is still retained.

$$\text{1 curie (Ci)} = 3.7 \times 10^{10} \text{ disintegrations per second (dps)}$$

$$= 2.22 \times 10^{12} \text{ disintegrations per minute (dpm)}$$

$$\text{1 millicurie (mCi)} = 3.7 \times 10^7 \text{ dps}$$

$$= 2.22 \times 10^9 \text{ dpm}$$

$$\text{1 microcurie } (\mu\text{Ci}) = 3.7 \times 10^4 \text{ dps}$$

$$= 2.22 \times 10^6 \text{ dpm}$$

The System Internationale (SI) unit for radioactivity is becquerel (Bq), which is defined as 1 dps. Thus,

$$\text{1 becquerel (Bq)} = 1 \text{ dps} = 2.7 \times 10^{-11} \text{Ci}$$

$$\text{1 kilobecquerel (kBq)} = 10^3 \text{ dps} = 2.7 \times 10^{-8} \text{Ci}$$

$$\text{2 megabecquerel (MBq)} = 10^6 \text{ dps} = 2.7 \times 10^{-5} \text{Ci}$$

$$\text{1 gigabecquerel (GBq)} = 10^9 \text{ dps} = 2.7 \times 10^{-2} \text{Ci}$$

$$\text{1 terabecquerel (TBq)} = 10^{12} \text{ dps} = 27 \text{ Ci}$$

Similarly,

$$\text{1 Ci} = 3.7 \times 10^{10} \text{ Bq} = 37 \text{ GBq}$$

$$\text{1 mCi} = 3.7 \times 10^7 \text{ Bq} = 37 \text{ MBq}$$

$$\text{1 } \mu\text{Ci} = 3.7 \times 10^4 \text{ Bq} = 37 \text{ kBq}$$

Specific Activity

The presence of "cold," or nonradioactive, atoms in a radioactive sample always induces competition between them in their chemical reactions or localization in a body organ, thereby compromising the concentration of the radioactive atoms in the organs. Thus, each radionuclide or radioactive sample is characterized by *specific activity*, which is defined as the radioactivity per unit mass of a radionuclide or a radioactive sample. For example, suppose that a 200-mg ^{123}I-labeled monoclonal antibody sample contains 350-mCi (12.95-GBq) ^{123}I radioactivity. Its specific activity would be $350/200 = 1.75$ mCi/mg or 64.75 MBq/mg. Sometimes, it is confused with concentration, which is defined as the radioactivity per unit volume of a sample. If a 10-ml radioactive sample contains 50 mCi (1.85 GBq), it will have a concentration of $50/10 = 5$ mCi/ml or 185 MBq/ml.

Specific activity is at times expressed as radioactivity per mole of a labeled compound, for example, mCi/mole (MBq/mole) or mCi/μmole (MBq/μmole) for ^{3}H-, ^{14}C-, and ^{35}S-labeled compounds.

The specific activity of a carrier-free radionuclide sample is related to its half-life: the shorter the half-life, the higher the specific activity. The specific activity of a carrier-free radionuclide with mass number A and half-life $t_{1/2}$ in hours can be calculated as follows:

Suppose 1 mg of a carrier-free radionuclide is present in the sample.

Number of atoms in the sample =

$$\frac{1 \times 10^{-3}}{A} \times 6.02 \times 10^{23} = \frac{6.02 \times 10^{20}}{A}$$

Decay constant $\lambda = \dfrac{0.693}{t_{1/2} \times 60 \times 60}$ sec^{-1}

Thus, disintegration rate $D = \lambda N$

$$= \frac{0.693 \times 6.02 \times 10^{20}}{t_{1/2} \times A \times 60 \times 60}$$

$$= \frac{1.1589 \times 10^{17}}{A \times t_{1/2}} \text{dps}$$

Thus, specific activity (mCi/mg) $= \dfrac{1.1589 \times 10^{17}}{A \times t_{1/2} \times 3.7 \times 10^7}$

$$= \frac{3.13 \times 10^9}{A \times t_{1/2}} \tag{3.15}$$

where A is the mass number of the radionuclide, and $t_{1/2}$ is the half-life of the radionuclide in hours.

From Eq. (3.15), specific activities of carrier-free ^{99m}Tc and ^{131}I can be calculated as 5.27×10^6 mCi/mg (1.95×10^5 GBq/mg) and 1.25×10^5 mCi/mg (4.6×10^3 GBq/mg), respectively.

Calculation

Some examples related to the calculation of radioactivity and its decay follow:

Problem 3.1
Calculate the total number of atoms and total mass of ^{201}Tl present in 10 mCi (370 MBq) of ^{201}Tl ($t_{1/2} = 3.04$ d).

Answer

For ^{201}Tl, $\lambda = \dfrac{0.693}{3.04 \times 24 \times 60 \times 60} = 2.638 \times 10^{-6} \ sec^{-1}$

$$D = 10 \times 3.7 \times 10^7 = 3.7 \times 10^8 \ dps$$

Using Eq. (3.2),

$$N = \frac{D}{\lambda} = \frac{3.7 \times 10^8}{2.638 \times 10^{-6}} = 1.40 \times 10^{14} \ atoms$$

Because 1 g·atom ^{201}Tl = 201 g ^{201}Tl = 6.02×10^{23} atoms of ^{201}Tl (Avogadro's number),

$$\text{Mass of } ^{201}\text{Tl in 10 mCi (370 MBq)} = \frac{1.40 \times 10^{14} \times 201}{6.02 \times 10^{23}}$$

$$= 46.7 \times 10^{-9} \ g$$

$$= 46.7 \ ng$$

Therefore, 10 mCi of ^{201}Tl contains 1.4×10^{14} atoms and 46.7 ng.

Problem 3.2
At 10:00 a.m., the ^{99m}Tc radioactivity was measured as 150 mCi (5.55 GBq) on Wednesday. What was the activity at 6 a.m. and 3 p.m. on the same day ($t_{1/2}$ of ^{99m}Tc = 6 hr)?

Answer
Time from 6 a.m. to 10 a.m. is 4 hr:

$$\lambda \text{ for } ^{99m}\text{Tc} = \frac{0.693}{6} = 0.1155 \ hr^{-1}$$

$$A_t = 150 \ mCi \ (5.55 \ GBq)$$

$$A_0 = ?$$

Using Eq. (3.6)

$$150 = A_0 e^{+0.1155 \times 4}$$

$$A_0 = 150 \times e^{0.462}$$

$$= 150 \times 1.5872$$

$$= 238.1 \ mCi \ (8.81 \ GBq) \ at \ 6 \ a.m.$$

Time from 10 a.m. to 3 p.m. is 5 hr:

$$A_0 = 150 \ mCi$$

$$A_t = ?$$

Using Eq. (3.6)

$$A_t = 150 \times e^{-0.1155 \times 5}$$

$$= 150 \times e^{-0.5775}$$

$$= 150 \times 0.5613$$

$$= 84.2 \text{ mCi (3.1 GBq) at 3 p.m.}$$

Problem 3.3

If a radionuclide decays at a rate of 30%/hr, what is its half-life?

Answer

$$\lambda = 0.3 \text{ hr}^{-1}$$

$$\lambda = \frac{0.693}{t_{1/2}}$$

$$t_{1/2} = \frac{0.693}{\lambda} = \frac{0.693}{0.3} \text{ hr} = 2.31 \text{ hr}$$

Problem 3.4

If 11% of ^{99m}Tc-labeled diisopropyliminodiacetic acid (DISIDA) is eliminated via renal excretion, 35% by fecal excretion, and 3.5% by perspiration in 5 hr from the human body, what is the effective half-life of the radiopharmaceutical ($T_p = 6$ hr for ^{99m}Tc)?

Answer

$$\text{Total biologic elimination} = 11\% + 35\% + 3.5\%$$

$$= 49.5\% \text{ in 5 hr}$$

Therefore, $T_b \approx 5$ hr

$$T_p = 6 \text{ hr}$$

$$T_e = \frac{T_b \times T_p}{T_b + T_p} = \frac{5 \times 6}{5 + 6} = \frac{30}{11} = 2.73 \text{ hr}$$

Successive Decay Equations

General Equation

In the preceding section, we derived equations for the disintegration rate or activity of any radionuclide that is decaying. Here we shall derive equations for the disintegration rate or activity of a radionuclide that is growing from another radionuclide and at the same time is itself decaying.

If a parent radionuclide p decays to a daughter radionuclide d, which in turn decays to another radionuclide (i.e., $p \rightarrow d \rightarrow$), then the rate of growth of d becomes

$$\frac{dN_d}{dt} = \lambda_p N_p - \lambda_d N_d \tag{3.16}$$

By integration, Eq. (3.16) becomes

$$(A_d)_t = \lambda_d N_d = \frac{\lambda_d (A_p)_0}{\lambda_d - \lambda_p}(e^{-\lambda_p t} - e^{-\lambda_d t}) \tag{3.17}$$

Equation (3.17) gives the disintegration rate of the daughter nuclide d at time t as a result of growth from the parent nuclide p and also is due to the decay of the daughter itself.

Transient Equilibrium

If $\lambda_d > \lambda_p$, that is, $(t_{1/2})_d < (t_{1/2})_p$, then $e^{-\lambda_d t}$ in Eq. (3.17) is negligible compared to $e^{-\lambda_p t}$ when t is sufficiently long. Then Eq. (3.17) becomes

$$(A_d)_t = \frac{\lambda_d (A_p)_0}{\lambda_d - \lambda_p}e^{-\lambda_p t}$$

$$= \frac{\lambda_d (A_p)_t}{\lambda_d - \lambda_p} \tag{3.18}$$

$$= \frac{(t_{1/2})_p (A_p)_t}{(t_{1/2})_p - (t_{1/2})_d} \tag{3.19}$$

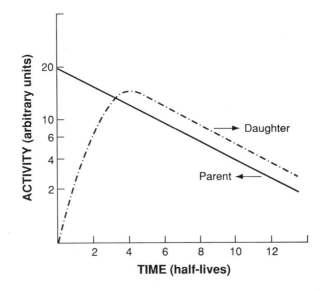

Fig. 3.4. Plot of activity versus time on a semilogarithmic graph illustrating the transient equilibrium. Note that the daughter activity reaches a maximum after about 4 to 5 half-lives, then reaches transient equilibrium and follows an apparent half-life of the parent. The daughter activity is higher than the parent activity at equilibrium.

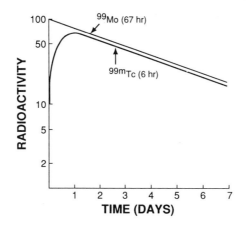

Fig. 3.5. Plot of logarithm of ^{99}Mo and ^{99m}Tc activities versus time showing transient equilibrium. The activity of the daughter ^{99m}Tc is less than that of the parent ^{99}Mo, because only 87% of ^{99}Mo decays to ^{99m}Tc. If 100% of the parent were to decay to the daughter, then the daughter activity would be higher than the parent activity after reaching equilibrium, as recognized from Eq. (3.19) and Figure 3.4.

This relationship is called the *transient equilibrium*. This equilibrium holds good when $(t_{1/2})_p$ and $(t_{1/2})_d$ differ by a factor of about 10 to 50. The semilogarithmic plot of this equilibrium equation is shown in Figure 3.4. The daughter nuclide initially builds up as a result of the decay of the parent nuclide, reaches a maximum, and then achieves the transient equilibrium decaying with an apparent half-life of the parent nuclide. It can be seen from Eq. (3.18) that the daughter activity is always greater than the parent activity, because $(t_{1/2})_p/([t_{1/2}]_p - [t_{1/2}]_d)$ is always greater than 1. In transient equilibrium, the maximum daughter activity is reached in about four half-lives of the daughter, while the equilibrium is achieved in almost seven to eight half-lives.

A typical example of transient equilibrium is ^{99}Mo ($t_{1/2} = 67$ hr) decaying to ^{99m}Tc ($t_{1/2} = 6$ hr). Because 87% of ^{99}Mo decays to ^{99m}Tc, and the remaining 13% to the ground state, Eqs. (3.17), (3.18), and (3.19) must be multiplied by a factor of 0.87. Therefore, in the time–activity plot, the ^{99m}Tc daughter activity will be lower than the ^{99}Mo parent activity (Fig. 3.5).

Secular Equilibrium

When $\lambda_d \gg \lambda_p$, that is, when the parent half-life is much longer than that of the daughter nuclide, in Eq. (3.15), we can neglect λ_p compared to λ_d. Then Eq. (3.15) becomes

$$(A_d)_t = (A_p)_t \tag{3.20}$$

Equation (3.20) is called the *secular equilibrium*. This equilibrium holds

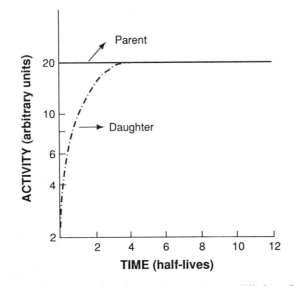

Fig. 3.6. Plot of activity versus time illustrating secular equilibrium. In equilibrium, the daughter activity becomes equal to that of the parent.

when the half-life of the parent is much longer than that of the daughter nuclide by more than a factor of 100 or so. In secular equilibrium, both parent and daughter activities are equal, and both decay with the half-life of the parent nuclide. A semilogarithmic plot of activity versus time representing secular equilibrium is shown in Figure 3.6. Typical examples of secular equilibrium are ^{113}Sn ($t_{1/2}$ = 117 days) decaying to ^{113m}In ($t_{1/2}$ = 100 min), and ^{68}Ge ($t_{1/2}$ = 280 days) decaying to ^{68}Ga ($t_{1/2}$ = 68 min).

Questions

1. Calculate (a) the total number of atoms and (b) the total mass of ^{131}I present in a 30-mCi (1.11 = GBq) ^{131}I sample ($t_{1/2}$ = 8.0 days).
2. Calculate (a) the disintegration rate per minute and (b) the activity in curies and becquerels present in 1 mg of ^{201}Tl ($t_{1/2}$ = 73 hr).
3. A radiopharmaceutical has a biologic half-life of 10 hr in humans and a physical half-life of 23 hr. What is the effective half-life of the radiopharmaceutical?
4. If the radioactivity of ^{111}In ($t_{1/2}$ = 2.8 days) is 200 mCi (7.4 GBq) on Monday noon, what is the activity (a) at 10:00 a.m. the Friday before and (b) at 1:00 p.m. the Wednesday after?
5. How long will it take for a 10-mCi (370-MBq) sample of ^{123}I ($t_{1/2}$ = 13.2 hr) and a 50-mCi (1.85-MBq) sample of ^{99m}Tc ($t_{1/2}$ = 6 hr) to possess the same activity?

6. What is the time interval during which ^{111}In ($t_{1/2}$ = 2.8 days) decays to 37% of the original activity?
7. For antibody labeling, 5 mCi of ^{111}InCl$_3$ is required. What amount of ^{111}InCl$_3$ should be shipped if transportation takes 1 day?
8. What are the specific conditions of transient equilibrium and secular equilibrium?
9. How long will it take for the decay of 7/8 of an ^{18}F ($t_{1/2}$ = 110 min) sample?
10. What fraction of the original activity of a radionuclide has decayed in a period equal to the mean life of the radionuclide?
11. A radioactive sample initially gives 8564 cpm and 2 hr later gives 3000 cpm. Calculate the half-life of the radionuclide.
12. If N atoms of a sample decay in one half-life, how many atoms would decay in the next half-life?
13. The ^{99}Mo ($t_{1/2}$ = 67 hr) and ^{99m}Tc ($t_{1/2}$ = 6 hr) are in transient equilibrium in a Moly generator. If 600 mCi (22.2 GBq) of ^{99}Mo is present in the generator, what would be the activity of ^{99m}Tc 132 hr later?
14. A radionuclide decays with a half-life of 10 days to a radionuclide whose half-life is 1.5 hr. Approximately how long will it take for the daughter to grow up to the parent activity?

Suggested Readings

Friedlander G, Kennedy JW, Miller JM. *Nuclear Chemistry and Radiochemistry.* 3rd ed. New York: Wiley; 1981.

Rollo FD, ed. *Nuclear Medicine Physics, Instrumentation and Agents.* St Louis: Mosby; 1977.

Saha GB. *Fundamentals of Nuclear Pharmacy.* 3rd ed. New York: Springer-Verlag; 1992.

Sorensen JA, Phelps ME. *Physics in Nuclear Medicine.* 2nd ed. New York: Grune & Stratton; 1987.

CHAPTER 4

Statistics of Radiation Counting

As mentioned in previous chapters, radioactive decay is a random process, and therefore one can expect fluctuations in the measurement of radioactivity. The detailed discussion of the statistical treatment of radioactive measurements is beyond the scope of this book. Only the salient points of statistics related to radiation counting are given here.

Error, Accuracy and Precision

In the measurement of any quantity, an error in or deviation from the true value is likely to occur. Errors can be two types: systematic and random. Systematic errors appear as constant deviations and arise from the malfunctioning instruments and inappropriate experimental conditions. These errors can be eliminated by rectifying the incorrect situations. Random errors are variable deviations and arise from the fluctuations in experimental conditions such as high-voltage fluctuations or statistical fluctuations in a quantity such as radioactive decay.

The accuracy of a measurement of a quantity indicates how closely it agrees with the "true" value. The precision of a series of measurements describes the reproducibility of the measurement and indicates the deviation from the "average" or "mean" value. The closer the measurement is to the average value, the higher the precision, whereas the closer the measurement is to the true value, the more accurate the measurement. Remember that a series of measurements may be quite precise, but their average value may be far from the true value (i.e., less accurate). Precision can be improved by eliminating the random errors, whereas better accuracy is obtained by removing both the random and systematic errors.

Mean and Standard Deviation

When a series of measurements is made on a radioactive sample, the most likely value of these measurements is the *average*, or *mean* value, which is obtained by adding the values of all measurements divided by the number of measurements. The mean value is denoted by $\bar{n}$.

The standard deviation of a series of measurements indicates the deviation from the mean value and is a measure of the precision of the measurements. Radioactive decay follows the Poisson distribution law, and from this, one can show that if a radioactive sample gives an average count of $\bar{n}$, then its standard deviation σ is given by

$$\sigma = \sqrt{\bar{n}} \tag{4.1}$$

The mean of measurements is then expressed as

$$\bar{n} \pm \sigma$$

If a series of measurements are made repeatedly on a radioactive sample giving a mean count $\bar{n}$, then the distribution of counts would normally follow a Poisson distribution. If the number of measurements is large, the distribution can be approximated by a Gaussian distribution, illustrated in Figure 4.1. It can be seen that 68% of all measurements fall within one standard deviation on either side of the mean, that is, within the range $\bar{n} \pm \sigma$; 95% of all measurements fall within the range $\bar{n} \pm 2\sigma$; and 99% fall within the range $\bar{n} \pm 3\sigma$.

The standard deviations in radioactive measurements indicate the statistical fluctuations of radioactive decay. For practical reasons, only single counts are obtained on radioactive samples instead of multiple repeat counts to determine the mean value. In this situation, if a single count n of a radioactive sample is large, then n can be estimated as close to $\bar{n}$; that is, $\bar{n} = n$ and $\sigma = \sqrt{n}$. It can then be said that there is a 68% chance that the true value of the count falls within $n \pm \sigma$ or that the count n falls within one standard deviation of the true value (Fig. 4.1). This is called the *68% confidence level*. That is, one is 68% confident that the count n is within one standard deviation of the true value. Similarly, *95% and 99% confidence levels* can be set at

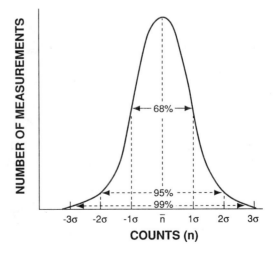

Fig. 4.1. A Gaussian distribution of radioactive measurements. Note the 68% confidence level at $\pm 1\sigma$, 95% confidence level at $\pm 2\sigma$, and 99% confidence level at $\pm 3\sigma$.

two standard deviations (2σ) and three standard deviations (3σ), respectively, of any single radioactive count.

Another useful quantity in the statistical analysis of the counting data is the percent standard deviation, which is given as

$$\%\sigma = \frac{\sigma}{n} \times 100 = \frac{100\sqrt{n}}{n} = \frac{100}{\sqrt{n}} \qquad (4.2)$$

Equation (4.2) indicates that as n increases, the $\%\sigma$ decreases, and hence, precision of the measurement increases. Thus, the precision of a count of a radioactive sample can be increased by accumulating a large number of counts in a single measurement. For example, for a count of 10,000, $\%\sigma$ is 1%, whereas for 1,000,000, $\%\sigma$ is 0.1%.

Problem 4.1
How many counts should be collected for a radioactive sample to have a 2% error at a 95% confidence level?

Answer
95% confidence level is 2σ, that is, $2\sqrt{n}$

$$2\% = \frac{2\sigma \times 100}{n} = \frac{2\sqrt{n} \times 100}{n}$$

Therefore,

$$2 = \frac{200}{\sqrt{n}}$$
$$\sqrt{n} = 100$$
$$n = 10,000 \text{ counts}$$

Standard Deviation of Count Rates

The standard deviation of a count rate is

$$\sigma_c = \frac{\sigma}{t}$$

where σ is the standard deviation of the total count n of a radioactive sample obtained in time t. Because n is equal to the count rate c times the counting time t,

$$\sigma_c = \sqrt{n}/t = \sqrt{ct}/t = \sqrt{\frac{c}{t}} \qquad (4.3)$$

Problem 4.2
A radioactive sample is counted for 12 min and gives 8640 counts. Calculate the count rate and its standard deviation for the sample.

Answer

$$\text{Count rate } c = \frac{8640}{12} = 720 \text{ counts per minute (cpm)}$$

Standard deviation,

$$\sigma_c = \sqrt{c/t} = \sqrt{720/12} \approx 8$$

Therefore, the count rate is 720 ± 8 cpm.

Propagation of Errors

Situations arise in which two quantities, say x and y, with their respective standard deviations, σ_x and σ_y, are either added, subtracted, multiplied, or divided. The standard deviations of the results of these arithmetic operations are given by the following expressions:

Addition:

$$\sigma_{x+y} = \sqrt{\sigma_x{}^2 + \sigma_y{}^2} \tag{4.4}$$

Subtraction:

$$\sigma_{x-y} = \sqrt{\sigma_x{}^2 + \sigma_y{}^2} \tag{4.5}$$

Multiplication:

$$\sigma_{(x \times y)} = (x \times y)\sqrt{(\sigma_x/x)^2 + (\sigma_y/y)^2} \tag{4.6}$$

Division:

$$\sigma_{(x/y)} = (x/y)\sqrt{(\sigma_x/x)^2 + (\sigma_y/y)^2} \tag{4.7}$$

Problem 4.3
A radioactive sample and the background were counted for 5 min and found to give 8000 counts and 3000 counts, respectively. Calculate the net count rate, its standard deviation, and percent standard deviation.

Answer

$$\text{Gross sample count rate} = \frac{8000}{5} = 1600 \text{ cpm}$$

$$\text{Background count rate} = \frac{3000}{5} = 600 \text{ cpm}$$

$$\text{Net count rate} = 1600 - 600 = 1000 \text{ cpm}$$

Using Eqs. (4.3) and (4.4)

$$\sigma_c = \sqrt{\left(\sqrt{\frac{1600}{4}}\right)^2 + \left(\sqrt{\frac{600}{4}}\right)^2}$$

$$= \sqrt{\frac{1600}{4} + \frac{600}{4}}$$

$$= \frac{1}{\sqrt{4}}\sqrt{1600 + 600}$$

$$= \frac{1}{2}\sqrt{2200}$$

$$= \frac{47}{2}$$

$$\approx 24 \text{ cpm}$$

Thus, the count rate of the sample is 1000 ± 24 cpm

$$\% \text{ standard deviation of the count rate} = \frac{24}{1000} \times 100 = 2.4\%$$

Problem 4.4
A thyroid patient is given a ^{131}I–NaI capsule to measure the 24-hr thyroid uptake. The 2-min counts are: standard, 90,000; room background, 1000; thyroid, 40,000; and thigh, 2000. Calculate the thyroid uptake and its percent standard deviation.

Answer
Net standard count $= 90,000 - 1000 = 89,000$

$$\sigma_s = \sqrt{90,000 + 1000}$$

$$= 302$$

Net thyroid count $= 40,000 - 2000 = 38,000$

$$\sigma_t = \sqrt{40,000 + 2000}$$

$$= 205$$

Percent thyroid uptake

$$= \frac{38,000}{89,000} \times 100 = 42.7\%$$

Percent standard deviation in uptake [using Eq. (4.7)]

$$= \frac{38,000}{89,000} \times \sqrt{\left(\frac{302}{89,000}\right)^2 + \left(\frac{205}{38,000}\right)^2} \times 100$$

$$= 0.427 \times \sqrt{0.000011514 + 0.000029103} \times 100$$

$$= 0.427 \times 0.006373 \times 100$$

$$= 0.27$$

Thus, the thyroid uptake $= 42.7 \pm 0.27\%$.

It should be noted that although all counts were taken for 2 min, count rates (cpm) were not used in the calculations. One can do similar calculations using count rates and obtain the same results.

Questions

1. Define *accuracy* and *precision*.
2. Do systematic errors give an accurate measurement? Can systematic errors give a precise measurement?
3. A radioactive sample gives 15,360 counts in 9 min:
   (a) What are the count rate of the sample and its standard deviation?
   (b) If the sample contained a background count rate of 60 cpm obtained from a 2-min count, what would be the net count rate of the sample and its standard deviation?
4. How many counts of a sample are to be collected to have a 1% error at the 95% confidence level?
5. Within how many standard deviations of a mean count of 62,001 is 730?
6. To achieve an estimated percent standard error of 3%, how many counts must collected?

Suggested Readings

Bahn AK. *Basic Medical Statistics.* New York: Grune & Stratton; 1972.

Martin PM. Nuclear medicine statistics. In: Rollo FD, ed. *Nuclear Medicine Physics, Instrumentation and Agents.* St. Louis: Mosby; 1977, pp 479–512.

CHAPTER 5

Production of Radionuclides

Nearly 3000 nuclides are known, of which approximately 2700 are radioactive, and the rest are stable. The majority of radionuclides are artificially produced in the cyclotron and reactor. Some short-lived radionuclides are available from the so-called radionuclide generators in which long-lived parents are loaded and decay to short-lived daughters. The following is a brief description of these sources of radionuclides.

Cyclotron-Produced Radionuclides

In a cyclotron (Fig. 5.1), charged particles (S) such as protons, deuterons, α particles, ^{3}He particles, and so forth are accelerated in circular paths within the dees (A and B) under vacuum by means of an electromagnetic field. These accelerated particles can possess a few kiloelectron volts (keV) to several billion electron volts (BeV) of kinetic energy depending on the design of the cyclotron. Because charged particles move along the circular paths under the magnetic field with gradually increasing energy, the larger the radius of the particle trajectory, the higher the kinetic energy of the particle. The charged particles are deflected by a deflector (D) through a window (W) outside the cyclotron to form an external beam.

When targets of stable elements are irradiated by placing them in the external beam of the accelerated particles or in the internal beam at a given radius inside a cyclotron, the accelerated particles interact with the target nuclei, and nuclear reactions take place. In a nuclear reaction, the incident particle may leave the nucleus after interaction, leaving some of its energy in it, or it may be completely absorbed by the nucleus, depending on the energy of the incident particle. In either case, a nucleus with excitation energy is formed and the excitation energy is disposed of by the emission of nucleons (i.e., protons and neutrons). Particle emission is followed by γ-ray emission when the former is no longer energetically feasible. Depending on the energy deposited by the incident particle, several nucleons are emitted randomly

Fig. 5.1. Schematics of a cyclotron. A and B, dees with vacuum; D, deflector; S, ion source; V, alternating voltage; W, window.

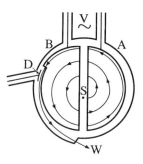

from the irradiated target nucleus, leading to the formation of different nuclides. As the energy of the irradiating particle is increased, more nucleons are emitted, and therefore a much wider variety of nuclides is produced.

Medical cyclotrons are compact cyclotrons that are used to produce routinely short-lived radionuclides, particularly those used in positron emission tomography. In these cyclotrons, protons, deuterons, and α particles of low-to-medium energy are available. These units are available commercially and can be installed in a relatively small space.

An example of a typical cyclotron-produced radionuclide is ^{111}In, which is produced by irradiating ^{111}Cd with 12-MeV protons in a cyclotron. The nuclear reaction is written as follows:

$$^{111}Cd(p, n)^{111}In$$

where ^{111}Cd is the target nuclide, the proton p is the irradiating particle, the neutron n is the emitted particle, and ^{111}In is the product radionuclide. In this case, a second nucleon may not be emitted, because there may not be enough energy left after the emission of the first neutron. The excitation energy that is not sufficient to emit any more nucleons will be dissipated by γ-ray emission.

As can be understood, radionuclides produced with atomic numbers different from those of the target isotopes do not contain any stable ("cold," or "carrier") isotope detectable by ordinary analytical methods, and such preparations are called *carrier-free*. In practice, however, it is impossible to have these preparations without the presence of any stable isotopes. Another term for these preparations is *no-carrier-added* (NCA), meaning that no stable isotope has been added purposely to the preparations.

The target material for irradiation must be pure and preferably mono-isotopic or at least enriched isotopically to avoid the production of extraneous radionuclides. Because various isotopes of different elements may be produced in a target, it is necessary to isolate isotopes of a single element; this can be accomplished by appropriate chemical methods such as solvent extraction, precipitation, ion exchange, and distillation. Cyclotron-produced radionuclides are usually neutron deficient and therefore decay by β^+ emission or electron capture.

Reactor-Produced Radionuclides

A variety of radionuclides is produced in nuclear reactors. A nuclear reactor is constructed with fuel rods made of fissile materials such as enriched ^{235}U and ^{239}Pu. These fuel nuclei undergo spontaneous fission with extremely low probability. *Fission* is defined as the breakup of a heavy nucleus into two fragments of approximately equal mass, accompanied by the emission of two to three neutrons with mean energies of about 1.5 MeV. In each fission, there is a concomitant energy release of ~ 200 MeV that appears as heat and is usually removed by heat exchangers to produce electricity in the nuclear power plant.

Neutrons emitted in each fission can cause further fission of other fissionable nuclei in the fuel rod, provided the right conditions exist. This obviously will initiate a chain reaction, ultimately leading to an explosive situation in the reactor. This chain reaction must be controlled, which is accomplished by the proper size, shape, and mass of the fuel material and other complicated and ingenious engineering techniques. To control a self-sustained chain reaction, excess neutrons (more than one) are removed by positioning cadmium rods in the fuel core (cadmium has a high probability of absorbing a thermal neutron).

The fuel rods of fissile materials are interspersed in the reactor core with spaces in between. Neutrons emitted with a mean energy of 1.5 MeV from the surface of the fuel rod have a low probability of interacting with other nuclei and therefore do not serve any useful purpose. It has been found, however, that neutrons with thermal energy (0.025 eV) interact with many other stable nuclei efficiently, producing various radionuclides. To make the high-energy neutrons, or so-called fast neutrons, more useful, they are thermalized or slowed down by interaction with low molecular weight materials, such as heavy water (D_2O), beryllium, and graphite (C), which are distributed in the spaces between the fuel rods. These materials are called *moderators*. The flux, or intensity, of the thermal neutrons so obtained ranges from 10^{11} to 10^{14} neutrons/cm$^2 \cdot$ sec, and they are useful in the production of many radionuclides. When a target element is inserted in the reactor core, a thermal neutron will interact with the target nucleus, with a definite probability of producing another nuclide. The probability of formation of a radionuclide by thermal neutrons varies from element to element.

In the reactor, two types of interaction with thermal neutrons are of importance in the production of useful radionuclides: fission of heavy elements and neutron capture or (n, γ) reaction. These two nuclear reactions are described next.

Fission or (n, f) Reaction

When a target of heavy elements is inserted in the reactor core, heavy nuclei absorb thermal neutrons and undergo fission. Fissionable heavy elements are ^{235}U, ^{239}Pu, ^{237}Np, ^{233}U, ^{232}Th, and many others having atomic numbers greater than 92. Fission of heavy elements may also be induced in a cyclotron by irradiation with high-energy charged particles, but fission probability depends on the type and energy of the irradiating particle. Nuclides produced by fission may range in atomic number from about 28 to nearly 65. These isotopes of different elements are separated by appropriate chemical procedures that involve precipitation, solvent extraction, ion exchange, chromatography, and distillation. The fission radionuclides are normally carrier-free or NCA, and therefore isotopes of high specific activity are available from fission. The fission products are usually neutron rich and decay by β^--emission.

Many clinically useful radionuclides such as ^{131}I, ^{99}Mo, ^{133}Xe, and ^{137}Cs are produced by fission of ^{235}U. An example of thermal fission of ^{235}U follows, showing a few representative radionuclides:

$$^{235}_{92}U + ^{1}_{0}n \rightarrow ^{236}_{92}U \rightarrow ^{131}_{53}I + ^{102}_{39}Y + 3^{1}_{0}n$$

$$\rightarrow ^{99}_{42}Mo + ^{135}_{50}Sn + 2^{1}_{0}n$$

$$\rightarrow ^{117}_{46}Pd + ^{117}_{46}Pd + 2^{1}_{0}n$$

$$\rightarrow ^{133}_{54}Xe + ^{101}_{38}Sr + 2^{1}_{0}n$$

$$\rightarrow ^{137}_{55}Cs + ^{97}_{37}Rb + 2^{1}_{0}n$$

$$\rightarrow ^{155}_{62}Sm + ^{78}_{30}Zn + 3^{1}_{0}n$$

$$\rightarrow ^{156}_{62}Sm + ^{77}_{30}Zn + 3^{1}_{0}n$$

Many other nuclides besides those mentioned in the example are also produced.

Neutron Capture or (n, γ) Reaction

In neutron capture reaction, the target nucleus captures one thermal neutron and emits γ-rays to produce an isotope of the same element. The radionuclide so produced is therefore not carrier-free, and its specific activity is relatively low. This reaction takes place in almost all elements with varying probability. Some examples of neutron capture reactions are ^{98}Mo(n, γ)^{99}Mo, ^{196}Hg(n, γ)^{197}Hg, and ^{50}Cr(n, γ)^{51}Cr. Molybdenum-99 so produced is called the *irradiated molybdenum* as opposed to the *fission molybdenum* described earlier. This method is commonly used in the analysis of trace elements in various samples.

The method of production and various characteristics of radionuclides commonly used in nuclear medicine are presented in Table 5.1.

Table 5.1. Characteristics of commonly used radionuclides.

Nuclide	Physical half-life	Mode of delay (%)	γ-ray energy* (MeV)	γ-ray abundance (%)	Common production method
$^{3}_{1}\text{H}$	12.3 yr	β^- (100)	—	—	$^{6}\text{Li}(n,\alpha)^{3}\text{H}$
$^{11}_{6}\text{C}$	20.4 min	β^+ (100)	0.511 (annihilation)	200	$^{10}\text{B}(d,n)^{11}\text{C}$ $^{14}\text{N}(p,\alpha)^{11}\text{C}$
$^{13}_{7}\text{N}$	10 min	β^+ (100)	0.511 (annihilation)	200	$^{12}\text{C}(d,n)^{13}\text{N}$ $^{16}\text{O}(p,\alpha)^{13}\text{N}$ $^{13}\text{C}(p,n)^{13}\text{N}$
$^{14}_{6}\text{C}$	5730 yr	β^- (100)	—	—	$^{14}\text{N}(n,p)^{14}\text{C}$
$^{15}_{8}\text{O}$	2 min	β^+ (100)	0.511 (annihilation)	200	$^{14}\text{N}(d,n)^{15}\text{O}$ $^{15}\text{N}(p,n)^{15}\text{O}$
$^{18}_{9}\text{F}$	110 min	β^+ (97) EC (3)	0.511 (annihilation)	194	$^{18}\text{O}(p,n)^{18}\text{F}$
$^{32}_{15}\text{P}$	4.3 day	β^- (100)	—	—	$^{32}\text{S}(n,p)^{32}\text{P}$
$^{57}_{27}\text{Co}$	271 days	EC (100)	0.014 0.122 0.136	9 86 11	$^{56}\text{Fe}(d,n)^{57}\text{Co}$
$^{67}_{31}\text{Ga}$	78 hr	EC (100)	0.093 0.184 0.300 0.393	40 20 17 5	$^{68}\text{Zn}(p,2n)^{67}\text{Ga}$
$^{68}_{31}\text{Ga}$	68 min	β^+ (89) EC (11)	0.511 (annihilation)	178	$^{68}\text{Zn}(p,n)^{68}\text{Ga}$
$^{99}_{42}\text{Mo}$	67 hr	β^- (100)	0.181 0.740 0.780	6 12 4	$^{98}\text{Mo}(n,\gamma)^{99}\text{Mo}$ $^{235}\text{U}(n,f)^{99}\text{Mo}$
$^{99m}_{43}\text{Tc}$	6.0 hr	IT (100)	0.140	90	$^{99}\text{Mo} \xrightarrow[67\,h]{\beta^-} {}^{99m}\text{Tc}$
$^{111}_{49}\text{In}$	2.8 days	EC (100)	0.171 0.245	90 94	$^{111}\text{Cd}(p,n)^{111}\text{In}$
$^{123}_{53}\text{I}$	13.2 hr	EC (100)	0.159	83	$^{121}\text{Sb}(\alpha,2n)^{123}\text{I}$ $^{127}\text{I}(p,5n)^{123}\text{Xe}$ $\downarrow$ 2.1 hr ^{123}I $^{124}\text{Xe}(p,2n)^{123}\text{Cs}$ $\downarrow$ 5.9 min ^{123}Xe $\downarrow$ 2.1 hr ^{123}I
$^{125}_{53}\text{I}$	60 days	EC (100)	0.035 X-ray (0.027–0.032)	7 140	$^{124}\text{Xe}(n,\gamma)^{125}\text{Xe}$ $^{125}\text{Xe} \xrightarrow[17\,hr]{EC} {}^{115}\text{I}$
$^{131}_{53}\text{I}$	8.0 days	β^- (100)	0.284 0.364 0.637	6 81 7	$^{130}\text{Te}(n,\gamma)^{131}\text{Te}$ $^{235}\text{U}(n,f)^{131}\text{Te}$ $^{131}\text{Te} \xrightarrow[25\,min]{\beta^-} {}^{131}\text{I}$ $^{235}\text{U}(n,f)^{131}\text{I}$
$^{133}_{54}\text{Xe}$	5.3 days	β^- (100)	0.081	37	$^{235}\text{U}(n,f)^{133}\text{Xe}$
$^{137}_{55}\text{Cs}$	30.0 yr	β^- (100)	0.662	85	$^{235}\text{U}(n,f)^{137}\text{Cs}$
$^{201}_{81}\text{Tl}$	73 hr	EC (100)	0.167 X-ray (0.069–0.083)	9.4 93	$^{203}\text{Tl}(p,3n)^{201}\text{Pb}$ $^{201}\text{Pb} \xrightarrow[9.3\,hr]{EC} {}^{201}\text{Tl}$

*γ-rays with abundance less than 4% have not been cited.
d, deuteron; EC, electron capture; f, fission; IT, isomeric transition; n, neutron; p, proton.
Data from Browne E, Finestone RB. *Table of Radioactive Isotopes*. New York: Wiley; 1986.

Target and Its Processing

Various types of targets have been designed and used for both reactor and cyclotron irradiation. In the design of targets, primary consideration is given to heat deposition in the target by irradiation with neutrons in the reactor or charged particles in the cyclotron. In both cases, the temperature can rise to 1000°C, and if proper material is not used or a method of heat dissipation is not properly designed, the target is likely to be burned or melted. For this reason, water cooling of the cyclotron probe to which the target is attached is commonly adopted. In the case of the reactor, the core cooling with heavy water is sufficient to cool the target. Most often, the targets are designed in the form of a foil to maximize heat dissipation.

The common form of the target is metallic foil; for example, copper, aluminum, uranium, vanadium, and so on. Other forms of targets are oxides, carbonates, nitrates, and chlorides contained in an aluminum tubing, which is then flattened to maximize the heat loss. Aluminum tubing is used because of its high melting point. In some cases, compounds are deposited on the appropriate metallic foil by vacuum distillation or by electrodeposition, and the plated foils are then used as targets. In specific cases, high-pressure gases (e.g. ^{124}Xe for ^{123}I production) and liquid targets (e.g. $H_2{}^{18}O$ for ^{18}F production) are also used.

Equation for Production of Radionuclides

While irradiating a target for the production of a radionuclide, it is essential to know various parameters affecting its production, preferably in a mathematical form, to estimate how much of it would be produced for a given set of parameters. These parameters are therefore discussed in some detail in a mathematical form.

The disintegration rate of a radionuclide produced by irradiation of a target material with charged particles in a cyclotron or with neutrons in a nuclear reactor is given by

$$D = IN\sigma(1 - e^{-\lambda t}) \tag{5.1}$$

where

D = disintegrations per second of the radionuclide produced

I = intensity or flux of the irradiating particles [number of particles/ $(cm^2 \cdot sec)$]

N = number of target atoms

σ = formation cross-section (probability) of the radionuclide (cm^2); it is given in units of "barn", which is equal to 10^{-24} cm^2

λ = decay constant given by $0.693/t_{1/2}$ (sec^{-1})

t = duration of irradiation (sec)

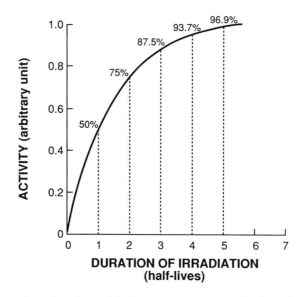

Fig. 5.2. Production of radionuclides in a reactor or a cyclotron. The activity produced reaches a maximum (saturation) in 5 to 6 half-lives of the radionuclide.

Equation (5.1) indicates that the amount of radioactivity produced depends on the intensity and energy (related to the cross-section σ) of the incident particles, the amount of the target material, the half-life of the radionuclide produced, and the duration of irradiation. The term $(1 - e^{-\lambda t})$ is called the *saturation factor* and approaches unity when t is approximately 5 to 6 half-lives of the radionuclide in question. At that time, the yield of the product nuclide becomes maximum, and its rates of production and decay become equal. For a period of irradiation of 5 to 6 half-lives, Eq. (5.1) becomes

$$D = IN\sigma \tag{5.2}$$

A graphic representation of Eqs. (5.1) and (5.2) is given in Figure 5.2.

The intensity of the irradiating particles is measured by various physical techniques, the description of which is beyond the scope of this book; however, the values are available from the operator of the cyclotron or the reactor. The formation cross-sections of various nuclides are determined by experimental methods using Eq. (5.1), and they have been compiled and published by many investigators. The number of atoms N of the target is calculated from the weight W of the material irradiated, the atomic weight A_w and natural abundance K of the target isotope, and Avogadro's number (6.02×10^{23}) as follows:

$$N = \frac{W \times K}{A_w} \times 6.02 \times 10^{23} \tag{5.3}$$

After irradiation, isotopes of different elements may be produced and therefore should be separated by the appropriate chemical methods. These radionuclides are identified and quantitated by detecting their radiations and measuring their half-lives by the use of the NaI(Tl) or Ge(Li) detectors coupled to a multichannel pulse height analyzer. They may also be assayed in an ionization chamber if the amount of radioactivity is high.

Radionuclide Generators

Radionuclide generators provide the convenient sources of short-lived radionuclides that are very useful clinically. The basic requirements for a generator are that a parent radionuclide has a longer half-life than that of the daughter, and the daughter can be easily separated from the parent. In a generator, a long-lived parent radionuclide is allowed to decay to its short-lived daughter radionuclide, and the latter is then chemically separated. The importance of radionuclide generators lies in the fact that they are easily transportable and serve as sources of short-lived radionuclides in institutions without cyclotron or reactor facilities.

A radionuclide generator consists of a glass or plastic column fitted at the bottom with a fretted disk. The column is filled with adsorbent material such as ion exchange resin, alumina, and so forth, on which the parent nuclide is adsorbed. The parent decays to the daughter until transient or secular equilibrium is established [Eqs. (3.19) and (3.20)] within several half-lives of the daughter. After equilibrium, the daughter appears to decay with the same half-life as the parent. Because of the differences in chemical properties, the daughter activity is eluted with an appropriate solvent, leaving the parent on the column. After elution, the daughter activity builds up again and can be eluted repeatedly.

A schematic diagram of a radionuclide generator is shown in Figure 5.3. The vial containing the eluant is first inverted onto needle A, and an evacuated vial is inverted on the other needle B. The vacuum in the vial on needle B draws the eluant from the vial A through the column and elutes the daughter nuclide, leaving the parent nuclide on the column. In some commercial generators, a bottle of eluant is placed inside the housing, and aliquots of eluant are used up in each elution by an evacuated vial.

An ideal radionuclide generator should be simple and sturdy for transportation. The generator eluate should be free of the parent nuclide and the adsorbent material.

Several radionuclide generators are available for ready supply of short-lived radionuclides: ^{99}Mo(67 hr)–^{99m}Tc(6 hr); ^{113}Sn(117 days)–^{113m}In(100 min); ^{68}Ge(271 days)–^{68}Ga(68 min); ^{82}Sr(25 days)–^{82}Rb(75 sec); ^{81}Rb(4.6 hr)–^{81m}Kr(13 sec). Of these, the ^{99}Mo–^{99m}Tc generator is the workhorse of nuclear pharmacy in nuclear medicine.

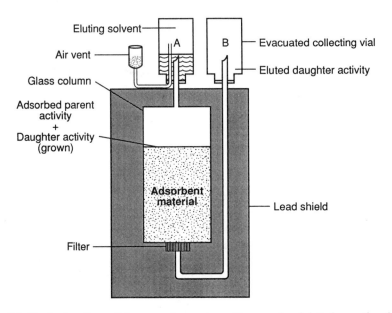

Fig. 5.3. Typical radionuclide generator system. Vacuum in vial B draws the eluant from vial A through adsorbent material, and the daughter is collected in vial B.

^{99}Mo–^{99m}Tc Generator

The ^{99}Mo–^{99m}Tc generator is constructed with alumina (Al_2O_3) loaded in a plastic or glass column. The ^{99}Mo activity is adsorbed on alumina in the chemical form $MoO_4{}^{2-}$ (molybdate) and in various amounts. The amount of alumina used is about 5–10 g depending on the ^{99}Mo activity. Currently, all generators use fission-produced ^{99}Mo. The growth and decay of ^{99m}Tc along with the decay of ^{99}Mo is shown in Figure 3.5 in (Chapter 3). The ^{99m}Tc activity is eluted with 0.9% NaCl solution (isotonic saline) in the chemical form of $Na^{99m}TcO_4$.

Considering that only 87% of ^{99}Mo decays to ^{99m}Tc, the ^{99m}Tc activity A_{Tc} can be obtained from Eq. (3.17) as follows:

$$A_{Tc} = 0.956(A_{Mo})_0(e^{-0.01034t} - e^{-0.1155t}) \qquad (5.4)$$

where $(A_{Mo})_0$ is the ^{99}Mo activity at $t = 0$, $\lambda_{Mo} = 0.01034$ hr^{-1}, and $\lambda_{Tc} = 0.1155$ hr^{-1}. The time t has the unit of hour. At transient equilibrium [from Eq. (3.18)],

$$A_{Tc} = 0.956(A_{Mo})_0 e^{-0.01034t}$$
$$= 0.956(A_{Mo})_t \qquad (5.5)$$

Upon elution with saline, approximately 75% to 85% of the total activity is eluted from the column. After about 4 half-lives, the ^{99m}Tc activity reaches maximum.

The ^{99}Mo activity is likely to be eluted in trace quantities along with ^{99m}Tc activity. This is called the ^{99}Mo, or *Moly, breakthrough*. According to the Nuclear Regulatory Commission (NRC) regulations, *the acceptable limit for ^{99}Mo breakthrough is 0.15 μCi (5.5 kBq) per millicurie (37 MBq) of ^{99m}Tc at the time of injection*. The ^{99}Mo breakthrough is determined by the detection of high-energy photons of ^{99}Mo in a dose calibrator after stopping 140-keV photons of ^{99m}Tc in a lead container (6-mm thick). The presence of ^{99}Mo gives unnecessary radiation dose to the patient.

Aluminum is also likely to be eluted with ^{99m}Tc activity and must be checked. Aluminum ion (Al^{3+}) is checked by the colorometric test using a paper strip impregnated with a coloring agent and comparing the intensity of the color developed by the sample solution with that by a standard test solution (10 μg/ml). Its permissible limit is 10 μg/ml of ^{99m}Tc solution. The presence of aluminum interferes with the preparation of ^{99m}Tc-labeled sulfur colloid by forming larger particles, which are trapped in the lungs. It also agglutinates the red blood cells during labeling.

Questions

1. Describe the different methods of production of radionuclides. Which method gives relatively more proton-rich and neutron-rich radionuclides?
2. What is the average number of neutrons emitted in fission?
3. What is the difference between carrier-free and no-carrier–added radionuclides?
4. Why are cadmium rods and graphite used in the reactor?
5. If ^{68}Zn is bombarded with protons in a cyclotron and three neutrons are emitted from the nucleus, what is the product of the nuclear reaction? Write the nuclear reaction.
6. (a) What are the primary requirements for a radionuclide generator?
   (b) How long does it take to reach transient equilibrium in a ^{99}Mo–^{99m}Tc generator?
   (c) What is the permissible limit of ^{99}Mo breakthrough in the ^{99m}Tc eluate?
   (d) A 20-mCi (740-MBq) ^{99m}Tc eluate is found to contain 10 μCi (0.37 MBq) of ^{99}Mo. Can this preparation be injected into a patient?
7. Why is Al^{3+} undesirable in the ^{99m}Tc eluate? What is the permissible limit of Al^{3+} in the ^{99m}Tc eluate?
8. A 2-Ci (74-GBq) ^{99}Mo–^{99m}Tc generator calibrated for Thursday noon was eluted at 10 a.m. on the following Monday. Calculate the ^{99m}Tc activity assuming 80% yield.
9. Calculate the activity in millicuries (MBq) of ^{111}In produced by irradiation of 1 g of pure ^{111}Cd for 3 hr with 12-MeV protons having a beam intensity of 10^{13} particles/($cm^2 \cdot$ sec). The cross section for formation of ^{111}In ($t_{1/2} = 2.8$ days) is 200 millibarns (1 millibarn = 10^{-27} cm^2).

Suggested Readings

Colombetti LG. Radionuclide generators. In: Rayudu GVS, ed. *Radiotracers for Medical Applications*. Boca Raton, Fla: CRC Press; 1983;II:133–168.

Friedlander G, Kennedy JW, Miller JM. *Nuclear and Radiochemistry*. 3rd ed. New York: Wiley; 1981.

Gelbard AS, Hara T, Tilbury RS, Laughlin JS. Recent aspects of cyclotron production of medically useful radionuclides. In: *Radiopharmaceuticals and Labelled Compounds*. Vienna: IAEA; 1973:239–247.

Noronha OPD, Sewatkar AB, Ganatra RD, et al. Fission-produced ^{99}Mo–^{99m}Tc generator system for medical use. *J Nucl Med Biol*. 1976;20:32–36.

Poggenburg JK. The nuclear reactor and its products. *Semin Nucl Med*. 1974;4:229–243.

Saha GB. Miscellaneous tracers for imaging. In: Rayudu GVS, ed. *Radiotracers for Medical Applications*. Boca Raton, Fla: CRC Press; 1983;II:119–132.

Saha GB. *Fundamentals of Nuclear Pharmacy*. 3rd ed. New York: Springer-Verlag; 1992.

Saha GB, MacIntyre WJ, Go RT. Cyclotron and positron emission tomography radiopharmaceuticals for clinical imaging. *Semin Nucl Med*. 1992; XXII: 150–161.

Tilbury RS, Laughlin JS. Cyclotron production of radioactive isotopes for medical use. *Semin Nucl Med*. 1974;4:245–255.

CHAPTER 6

Interaction of Radiation With Matter

All particulate and electromagnetic radiations may interact with the atoms of the absorber during their passage through it, producing ionization and excitation of the absorber atoms. These radations are called *ionizing radiations*. The mechanisms of interaction, however, differ for the two types of radiation, and therefore they are discussed separately.

Interaction of Charged Particles With Matter

The energetic charged particles such as α particles, protons, deuterons, and β particles (electrons) interact with the absorber atoms, while they pass through it. The interaction occurs primarily with the electrons of the absorber atoms and rarely with the nucleus. During the interaction, both ionization and excitation as well as the breakdown of the molecule may occur. In excitation, the charged particle transfers all or part of its energy to the orbital electrons, raising them to higher energy shells. In ionization, the energy transfer may be sufficient to overcome the binding energy of the orbital electrons, ultimately ejecting them from the atom. Electrons ejected from the atoms by the incident charged particles are called *primary electrons*, which may have sufficient kinetic energy to produce further excitation or ionization in the absorber. The high-energy secondary electrons from secondary ionizations are referred to as *delta* (δ-) rays. The process of excitation and ionization will continue until the incident particle and all electrons come to rest. Both these processes may rupture chemical bonds in the molecules of the absorber, forming various chemical entities.

In ionization, an average energy of W keV is required to produce an ion pair in the absorber and varies somewhat with the type of absorber. The value of W is about 35 eV in air and less in oxygen and xenon gases but falls in the range of 25 45 cV for most gases. The process of ionization, that is, the formation of ion pairs, is frequently used as a means of the detection of charged particles in ion chambers and Geiger–Müller counters described in Chapter 7.

Three important quantities associated with the passage of charged particles through matter are specific ionization, linear energy transfer, and range of the particle in the absorber, and these are described next.

Specific Ionization

Specific ionization (SI) is the total number of ion pairs produced per unit length of the path of the incident radiation. The SI values of α particles are slightly greater than those of protons and deuterons, which in turn are larger than those of electrons.

Specific ionization increases with decreasing energy of the charged particle because of the increased probability of interaction at low energies. Therefore, toward the end of the travel, the charged particle shows a sharp increase in ionization. This peak ionization is called *Bragg ionization*. This phenomenon is predominant for heavy charged particles, whereas it is negligible for electrons.

Linear Energy Transfer

The linear energy transfer (LET) is the amount of energy deposited per unit length of the path by the radiation. From the preceding, it is clear that

$$\text{LET} = \text{SI} \times W \tag{6.1}$$

The LET is expressed in units of $keV/\mu m$ and is very useful in concepts of radiation protection. Electromagnetic radiations and β particles interact with matter, losing only little energy per interaction and therefore have low LETs. In contrast, heavy particles (α particles, neutrons, and protons) lose energy very rapidly, producing many ionizations in a short distance, and thus have high LETs. Some comparative approximate LET values in $keV/\mu m$ in tissue are 0.5 for 3-MV x-rays, 3.0 for 250-keV x-rays, 100 for 5-MeV α particles, 20 for 14-MeV neutrons, 0.25 for 1-MeV electrons.

Problem 6.1
If a particulate radiation produces 45,000 ion pairs per centimeter in air, calculate the LET of the radiation.

Answer

$$W = 35 \text{ keV per ion pair}$$

Using Eq. (6.1),

$$\text{LET} = \text{SI} \times W$$

$$= 45,000 \times 35$$

$$= 1,575,000 \text{ keV/cm}$$

$$= 157.5 \text{ keV}/\mu m$$

Range

The range (R) of a charged particle in an absorber is the straight-line distance traversed by the particle in the direction of the particle. The range of a particle depends on the mass, charge, and kinetic energy of the particle and also on the density of the absorber. The heavier and more highly charged particles have shorter ranges than lighter and lower charged particles. The range of charged particles increases with the energy of the particle. Thus, a 10-MeV particle will have a longer range than a 1-MeV particle. The range of the particle depends on the density of the absorber, in that the denser the absorber, the shorter the range.

Depending on the type of the charged particle, the entire path of travel may be unidirectional along the initial direction of motion, or tortuous (Fig. 6.1). Because the α particle loses only a very small fraction of energy in a single collision with an electron as a result of its heavier mass and is not appreciably deflected in the collision, the α-particle path is nearly a straight line along its initial direction (Fig. 6.1a). In contrast, β particles or electrons interact with extranuclear orbital electrons of the same mass and are deflected considerably. This leads to tortuous paths of these particles (Fig. 6.1b). In this situation, the true range is less than the total path traveled by the particle.

It is seen that the ranges of all identical particles in a given absorber are not exactly the same but show a spread of 3% to 4% (Fig. 6.2). This phenomenon, referred to as the *straggling of the ranges*, results from the statistical fluctuations in the number of collisions and in the energy loss per collision. The range straggling is less prominent with α particles but is severe with electrons because it is mostly related to the mass of the particle. The light mass electrons are considerably deflected during collisions and hence exhibit more straggling. If the transmission of a beam of charged particles through absorbers of different thicknesses is measured, the beam intensity will remain constant until the region of range straggling is encountered, where the beam intensity falls sharply from its initial value to zero. The absorber thickness

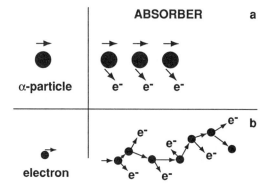

Fig. 6.1. Concept of passage of α particles and electrons through an absorber: (a) heavy α particles move in almost a straight line; (b) light electrons move in zigzag paths.

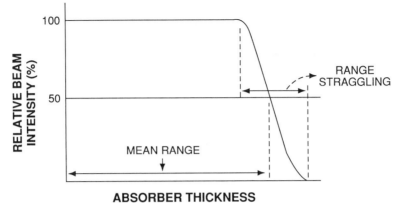

Fig. 6.2. Mean range and straggling of charged particles in an absorber.

that reduces the beam intensity by one half is called the *mean range*. The mean range of heavier particles such as α particles is more well defined than that of electrons. Because β^- particles are emitted with a continuous energy spectrum, their absorption, and hence their ranges, become quite complicated.

Bremsstrahlung

When energetic charged particles, particularly electrons, pass through matter and come close to the nucleus of the atom, they lose energy as a result of deceleration in the Coulomb field of atomic nuclei. The loss in energy appears as an x-ray which is called *bremsstrahlung* (German for "braking" or "slowing down" radiation). These bremsstrahlung radiations are commonly used in radiographic procedures and are generated by striking a tungsten target with a highly accelerated electron beam.

Bremsstrahlung production increases with the kinetic energy of the particle and the atomic number (Z) of the absorber. For example, a 10-MeV electron loses about 50% of its energy by bremsstrahlung, whereas a 90-MeV electron loses almost 90% of its energy by this process. Also, bremsstrahlung is unimportant in air, aluminum, and so forth, whereas it is very significant in heavy metals such as lead and tungsten. The bremsstrahlung production is proportional to Z^2 of the absorber atom. High-energy β^- particles from radionuclides such as ^{32}P can produce bremsstrahlung in heavy metals such as lead and tungsten. For this reason, these radionuclides are stored in low-Z materials such as plastic containers rather than shielding directly by lead.

Bremsstrahlung is inversely proportional to the mass of the charged particles and therefore is insignificant for heavy particles, namely α particles, protons, and so forth, because the probability of penetrating close to the nuclei is extremely low due to their larger masses.

Annihilation

When energetic β^+ particles pass through an absorber, they lose energy via interaction with orbital electrons of the atoms of the absorber. When the β^+ particle comes to almost rest after losing all energy, it combines with an orbital electron of the absorber atom and produces two 511-keV annihilation radiations that are emitted in exactly opposite directions. These annihilation radiations are the basis of positron emission tomography (PET) in which two photons are detected in coincidence.

Interaction of γ-Radiations With Matter

Mechanism of Interaction of γ-Radiations

When γ-rays pass through matter, they interact with the orbital electrons or the nucleus of the absorber atom, whereby their energy is lost. Unlike particles, the γ-ray photons can lose all of their energy, or a fraction of it, in a single encounter. The specific ionization of γ-rays is one-tenth to one-hundredth of that caused by an electron of the same energy. There is no quantity equivalent to a range of particles for γ-rays, but they travel a long path in the absorber before losing all energy. Therefore, the γ-rays are referred to as *penetrating radiations*. The average energy loss per ion pair produced by the photons is the same as for electrons, that is, 35 keV in air.

There are three mechanisms by which γ-rays interact with absorber atoms during their passage through matter, and they are described next.

Photoelectric Effect

In the photoelectric effect, the incident γ-ray transfers all its energy to an orbital electron of the absorber atom whereby the electron, called the *photoelectron*, is ejected with kinetic energy equal to $E_\gamma - E_B$, where E_γ and E_B are the energy of the γ-ray and the binding energy of the electron, respectively (Fig. 6.3). The photoelectron loses its energy by ionization and excitation in the absorber, as discussed previously. The photoelectric effect occurs pri-

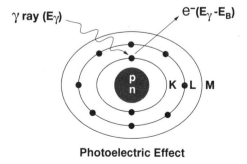

Fig. 6.3. The photoelectric effect in which a γ-ray with energy E_γ transfers all its energy to a K-shell electron, and the electron is ejected with $E_\gamma - E_B$, where E_B is the binding energy of the K-shell electron.

marily in the low-energy range and decreases sharply with increasing photon energy. It also increases very rapidly with increasing atomic number of the absorber atom. Roughly, the photoelectric effect is proportional to Z^5/E_γ^3. The photoelectric contribution from the 0.15-MeV γ-rays in aluminum ($Z = 13$) is about the same ($\sim 5\%$) as that from the 4.7-MeV γ-rays in lead ($Z = 82$).

The photoelectric effect occurs primarily with the K-shell electrons, with about 20% contribution from the L-shell electrons and even less from higher shells. There are sharp increases (discontinuities) in photoelectric effects at energies exactly equal to binding energies of K-, L- (etc.) shell electrons. These are called K-, L- (etc.) absorption edges. The vacancy created by the ejection of an orbital electron is filled in by the transition of an electron from the upper energy shell. It is then followed by emission of a characteristic x-ray or Auger electron, analogous to the situations in internal conversion or electron capture decay.

Compton Scattering

In Compton scattering, the γ-ray photon transfers only a part of its energy to an electron mostly belonging to an outer shell of the absorber atom, and the electron is ejected. The photon, itself with less energy, is deflected from its original direction (Fig. 6.4). This process is called the *Compton scattering*. The scattered photon of lower energy may then undergo further photoelectric or Compton interaction, and the Compton electron may cause ionization or excitation, as discussed previously.

At low energies, only a small fraction of the photon energy is transferred to the Compton electron, and the photon may be scattered in either the forward

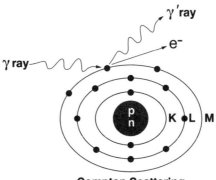

Compton Scattering

Fig. 6.4. The Compton scattering, in which a γ-ray interacts with an outer orbital electron of an absorber atom. Only a part of the photon energy is transferred to the electron, and the photon itself is scattered at an angle. The scattered photon may undergo subsequent photoelectric effect or Compton scattering in the absorber or may escape the absorber.

or the backward direction. If the photon is backscattered, that is, scattered at 180°, then the backscattered photon has the energy E_{sc} given by the expression:

$$E_{sc} = E_\gamma/(1 + E_\gamma/256) \qquad (6.2)$$

where E_γ is the initial photon energy in keV. It can be seen that as the photon energy increases, the scattered photon energy approaches a maximum of 256 keV, and the remaining energy is given to the Compton electron. Thus, at higher energies, more energy is transferred to the Compton electron, and both the scattered photon and the Compton electron are predominantly scattered in the forward direction.

Compton scattering is independent of the atomic number Z of the absorber. For photon energies above 0.5 MeV, it is almost inversely proportional to E_γ; that is, it decreases linearly with increasing photon energy. Compton scattering contributes primarily in the energy range of 0.1 to 10 MeV, depending on the type of absorber.

Pair Production

When the γ-ray photon energy is greater than 1.02 MeV, the photon interacts with the nucleus of the absorber atom during its passage through it, and a positive electron and a negative electron are produced at the expense of the photon (Fig. 6.5). The energy in excess of 1.02 MeV appears as the kinetic energy of the two particles. This process is called *pair production*. It varies almost linearly with Z^2 of the absorber and increases slowly with the energy of the photon. In soft tissue, pair production is insignificant at energies up to 10 MeV above 1.02 MeV. Positive electrons created by pair production are

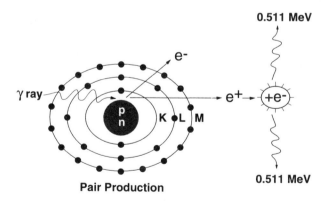

Pair Production

Fig. 6.5. Illustration of the pair production process. An energetic γ-ray with energy greater than 1.02 MeV interacts with the nucleus, and one positive electron (e⁺) and one negative electron (e⁻) are produced at the expense of the photon. The photon energy in excess of 1.02 MeV appears as the kinetic energy of the two particles. The positive electron eventually undergoes annihilation to produce two 511-keV photons emitted in opposite directions.

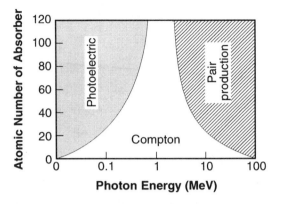

Fig. 6.6. Relative contributions of the photoelectric effect, Compton scattering, and pair production as a function of photon energy in absorbers of different atomic numbers. (Adapted with permission from Hendee WR. *Medical Radiation Physics.* 1st ed. Chicago: Year Book Medical Publishers, Inc; 1970:141.)

annihilated to produce two 0.511-MeV photons identical to those produced by positrons from radioactive decay.

The relative importance of photoelectric, Compton, and pair production interactions with absorbers of different atomic numbers is shown in Figure 6.6, as a function of the energy of the incident photons. It is seen that Compton scattering is the predominant mode of interaction in body tissues ($Z \approx 20$) for photons of 0.15–1.0 MeV, whereas in high-Z materials, the photoelectric effect is the primary mode in this energy range.

Photodisintegration

When the γ-ray energy is very high (> 10 MeV), the photon may interact with the nucleus of the absorber atom and transfer sufficient energy to the nucleus such that one or more nucleons may be emitted. This process is called the *photodisintegration reaction,* or *photonuclear reaction* and produces new nuclides. The (γ, n) reactions on targets such as ^{12}C and ^{14}N have been used to produce ^{11}C and ^{13}N radionuclides but now are rarely used to produce radionuclides.

Attenuation of γ-Radiations

Linear and Mass Attenuation Coefficients

γ-ray and x-ray photons are either attenuated or transmitted as they travel through an absorber. Attenuation results from absorption by the photoelectric effect, Compton scattering, and pair production at higher energies. Depending on the photon energy and the density and thickness of the absorber, some of the photons may pass through the absorber without any interaction

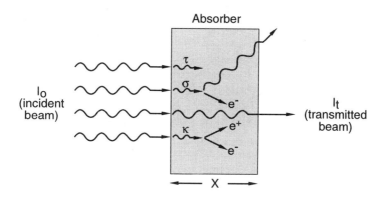

Fig. 6.7. Illustration of attenuation of a photon beam (I_0) in an absorber of thickness x. Attenuation comprises a photoelectric effect (τ), Compton scattering (σ), and pair production (κ). Photons passing through the absorber without interaction constitute the transmitted beam (I_t).

leading to the transmission of the photons (Fig. 6.7). Attenuation of γ-radiations is an important factor in radiation protection.

As shown in Figure 6.7, if a photon beam of initial intensity I_0 passes through an absorber of thickness x, then the transmitted beam I_t is given by the exponential equation

$$I_t = I_0 e^{-\mu x} \qquad (6.3)$$

where μ is the linear attenuation coefficient of the absorber for the photons of interest and has the unit of cm^{-1}. The factor $e^{-\mu x}$ represents the fraction of the photons transmitted. Because attenuation is primarily due to photoelectric, Compton, and pair production interactions, the linear attenuation coefficient μ is the sum of photoelectric coefficient (τ), Compton coefficient (σ), and pair production coefficient (κ). Thus,

$$\mu = \tau + \sigma + \kappa \qquad (6.4)$$

Linear attenuation coefficients normally decrease with the energy of the γ-ray or x-ray photons and increase with the atomic number and density of the absorber. The relative contributions of photoelectric effect, Compton scattering, and pair production in water (equivalent to body tissue) at different energies are illustrated in Figure 6.8.

An important quantity, μ_m, called the *mass attenuation coefficient*, is given by the linear attenuation coefficient divided by the density ρ of the absorber

$$\mu_m = \frac{\mu}{\rho} \qquad (6.5)$$

The mass attenuation coefficient μ_m has the unit of cm^2/g or cm^2/mg. The mass attenuation coefficients for fat, bone, muscle, iodine, and lead are given in Figure 6.9.

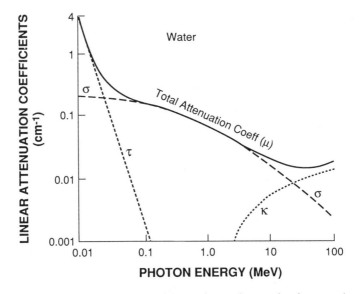

Fig. 6.8. Plot of linear attenuation coefficient of γ-ray interaction in water (equivalent to body tissue) as a function of photon energy. The relative contributions of photoelectric, Compton, and pair production processes are illustrated.

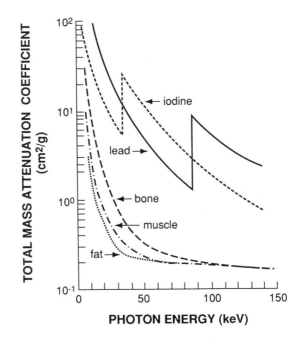

Fig. 6.9. Attenuation coefficients for fat, muscle, bone, iodine, and lead as a function of photon energy. (Adapted with permission from Hendee WR. *Medical Radiation Physics.* 1st ed. Chicago: Year Book Medical Publishers, Inc; 1970:221.)

Table 6.1. Half-value layer values (HVLs) of lead for commonly used radionuclides.*

Radionuclides	HVL, Lead (cm)
^{137}Cs	0.65
^{99m}Tc	0.03
^{201}Tl	0.02
^{99}Mo	0.70
^{67}Ga	0.10
^{123}I	0.04
^{111}In	0.10
^{125}I	0.003
^{57}Co	0.02
^{131}I	0.30
^{18}F	0.39

*Adapted from Goodwin PN. Radiation safety for patients and personnel. In: Freeman LM, ed. *Freeman and Johnson's Clinical Radionuclide Imaging.* 3rd ed. Philadelphia: WB Saunders Co; 1984:320.

Half-Value Layer

The concept of half-value layer (HVL) of an absorbing material for γ- or x-radiations is important in the design of shielding for radiation protection. It is defined as the thickness of the absorber that reduces the intensity of a photon beam by one-half. Thus, an HVL of an absorber around a source of γ-radiations with an exposure rate of 150 mR/hr will reduce the exposure rate to 75 mR/hr. The HVL depends on the energy of the radiation and the atomic number of the absorber. It is greater for high-energy photons and smaller for high-Z materials.

For monoenergetic photons, the HVL of an absorber is related to its linear attenuation coefficient as follows:

$$HVL = \frac{0.693}{\mu} \tag{6.6}$$

Because μ has the unit of cm^{-1}, the HVL has the unit of cm. The HVLs of lead for different radionuclides are given in Table 6.1.

Another important quantity, tenth-value layer (TVL), is the thickness of an absorber that reduces the initial beam by a factor of 10. It is given by

$$TVL = \frac{\ln(0.1)}{\mu}$$

$$= \frac{2.30}{\mu} \tag{6.7}$$

$$= 3.32 \ HVL \tag{6.8}$$

Problem 6.2

If the HVL of lead for the 140-keV photons of ^{99m}Tc is 0.03 cm of lead, calculate the linear attenuation coefficient of lead for the 140-keV photons and the amount of lead needed to reduce the exposure of a point source of radiation by 70%.

Answer

$$\mu = \frac{0.693}{\text{HVL}} = \frac{0.693}{0.03} = 23.1 \text{ cm}^{-1}$$

Because the initial beam is reduced by 70%, the remaining beam is 30%.

$$0.3 = 1 \times e^{-23.1 \times x}$$

$$\ln(0.3) = -23.1 \times x$$

$$1.20 = 23.1 \times x$$

$$x = 0.052 \text{ cm}$$

$$= 0.52 \text{ mm}$$

Thus, 0.52 mm of lead will reduce a beam of 140-keV photons by 70%.

Interaction of Neutrons With Matter

Because neutrons are neutral particles, their interactions in the absorber differ from those of the charged particles. They interact primarily with the nucleus of the absorber atom and very little with the atomic electrons. The neutrons can interact with the atomic nuclei in three ways: elastic scattering, inelastic scattering, and neutron capture. If the sum of the kinetic energies of the neutron and the nucleus before collision is equal to the sum of these quantities after collision, then the interaction is called *elastic*. If a part of the initial energy is used for the excitation of the struck nucleus, the collision is termed *inelastic*. In neutron capture, a neutron is captured by the absorber nucleus, and a new excited nuclide is formed. Depending on the energy deposited, an α particle, a proton, a neutron, or γ-rays can be emitted from the excited nucleus, and a new product nuclide is produced.

Questions

1. (a) What is the difference between excitation and ionization?
   (b) How are δ-rays produced?
   (c) How much energy is required on the average to produce an ion pair?
2. Define specific ionization (SI), linear energy transfer (LET), and range (R).
3. Electromagnetic radiations and electrons have low LETs compared to heavy particles (e.g., α particles), which have high LETs. Explain.

4. The range of an α particle is almost equal to the total path traveled, whereas the range of an electron is less than the total path traveled by the particle. Explain.

5. Indicate how the range of a charged particle is affected by the following conditions:
   (a) As the mass increases, the range increases or decreases.
   (b) As the energy of the particle increases, the range increases or decreases.
   (c) As the charge of the particle increases, the range increases or decreases.

6. Define Bragg ionization and straggling of ranges. Which has more straggling, an α particle or an electron? Explain.

7. How is bremsstrahlung produced? Does its production increase or decrease with increasing kinetic energy of the electron and the atomic number of the absorber? Explain why ^{32}P is stored in plastic containers.

8. Discuss the mechanism of the photoelectric effect. Does this process increase or decrease with increasing energy of the γ-ray and with increasing atomic number of the absorber?

9. A 0.495-MeV γ-ray interacts with a K-shell electron by the photoelectric process. If the binding energy of the K-shell electron is 28 keV, what happens to the rest of the photon energy?

10. (a) Explain the Compton scattering of electromagnetic radiations in the absorber.
    (b) Does it depend on the atomic number of the absorber?
    (c) How is it affected by the γ-ray energy?

11. If a relatively high-energy γ-ray is scattered at 180° (backscattered) by the Compton scattering, what is the maximum energy of the scattered photon?

12. (a) How does pair production occur?
    (b) Why does pair production require a minimum of 1.02 MeV for γ-ray energy?
    (c) Is this process affected by the atomic number of the absorber and the photon energy?

13. Which electrons of the absorber atom are involved in the photoelectric and Compton interactions of electromagnetic radiations?

14. (a) Discuss the attenuation of a photon beam passing through an absorber.
    (b) Does it depend on the density and the atomic number of the absorber?
    (c) Define the half-value layer (HVL) of an absorbing material for a γ-ray energy.

15. If 1 mCi of a radionuclide is adequately shielded by 5 HVLs of a shielding material, how many HVLs are needed to provide equal shielding for (a) 5 mCi and (b) 8 mCi?

16. A 1-mm lead apron will afford approximately twice as much protection as a 0.5-mm apron, or does this shielding depend on the energy of the radiation?

17. How many HVLs are approximately equivalent to three tenth-value layers?
18. Suppose 5% of the 364-keV photons of ^{131}I are transmitted after passing through a lead brick of 10-cm thickness. Calculate the HVL of lead for ^{131}I.
19. There is a 75% chance that a monoenergetic photon beam will be attenuated by 4 mm of lead. What is the HVL of lead for the photon?
20. Which of the following radiations has the highest LET?
    (a) 120-keV x-ray
    (b) 100-keV electron
    (c) 5-MeV α particle
    (d) 10-MeV proton
    (e) 14-MeV neutron

Suggested Readings

Friedlander G, Kennedy JW, Miller JM. *Nuclear and Radiochemistry*. 3rd ed. New York: Wiley; 1981.

Harvey BG. *Introduction to Nuclear Physics and Chemistry*. 2nd ed. Englewood Cliffs, NJ: Prentice-Hall; 1969.

Hendee WR. *Medical Radiation Physics*. 2nd ed. Chicago: Year Book Medical Publishers; 1979.

Johns HE, Cunningham JR. *The Physics of Radiology*. 4th ed. Springfield, Ill: Charles C Thomas; 1983.

Lapp RE, Andrews HL. *Nuclear Radiation Physics*. 4th ed. Englewood Cliffs, NJ: Prentice-Hall; 1972.

CHAPTER 7

Gas-Filled Detectors

Principles of Gas-Filled Detectors

The operation of a gas-filled detector is based on the ionization of gas molecules by radiation, followed by collection of the ion pairs as charge or current with the application of a voltage between two electrodes. The measured charge or current is proportional to the applied voltage and the amount of radiation, and depends on the type and pressure of the gas.

A schematic diagram of a gas-filled detector is shown in Figure 7.1. When an ionizing radiation beam passes through the gas, it will cause ionization of the gas molecules and ion pairs will be produced depending on the type and pressure of the gas. When a voltage is applied between the two electrodes, the negative electrons will move to the anode and the positive ions to the cathode, thus producing a current that can be measured on a meter.

At very low voltages, the ion pairs do not receive enough acceleration to reach the electrodes and therefore may combine together to form the original molecule instead of being collected by the electrodes. This region is called the *region of recombination* (Fig. 7.2). As the applied voltage is gradually increased, a *region of saturation* is encountered, where the current measured remains almost the same over the range of applied voltages. In this region, only the primary ion pairs formed by the initial radiations are collected. Individual events cannot be detected; only the total current passing through the chamber is measured. Because specific ionization differs for α-, β-, and γ-radiations, the amount of current produced by these radiations differs in this region. The voltage in this region is of the order of 50–300 V. Ionization chambers such as dose calibrators are operated in this region.

When the applied voltage is further increased, the electrons and positive ions gain such high velocities and energies during their acceleration toward the electrodes that they cause secondary ionization. The latter will increase the measured current. This process is referred to as the *gas amplification*. This factor can be as high as 10^6 per individual primary event depending on the design of the gas detector and the applied voltage. In this region, the total current measured is equal to the number of ionizations caused by the pri-

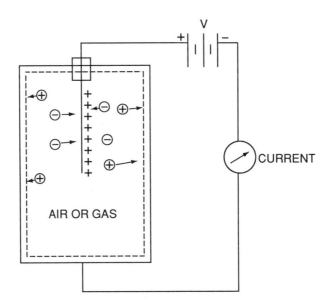

Fig. 7.1. A schematic diagram of a gas-filled detector illustrating the principles of operation.

mary radiation multiplied by the gas amplification factor. In this region, the current increases with the applied voltage in proportion to the initial number of ion pairs produced by the incident radiation. Therefore, as in the case of the region of saturation, the current amplification is relatively proportional for α-, β-, and γ-radiations. This region is referred to as the *proportional region* (see Fig. 7.2). Proportional counters are usually filled with 90% argon and 10% methane at atmospheric pressure. These counters can be used to count individual counts and to discriminate radiations of different energies. These counters, however, are not commonly used for γ- and x-ray counting because of poor counting efficiency ($<1\%$).

As the applied voltage is increased further, the current produced by different types of radiation tends to become identical. The voltage range over which the current tends to converge is referred to as the *region of limited proportionality*. This region is not practically used for detecting any radiation.

With additional increase in voltage beyond the region of limited proportionality, the current becomes identical, regardless of how many ion pairs are produced by the incident radiations. This region is referred to as the *Geiger region* (see Fig. 7.2). In the Geiger voltage region, the current is produced by an avalanche of interactions. When highly accelerated electrons strike the anode with a great force, ultraviolet (UV) light is emitted, which causes further emission of photoelectrons by gas ionization and from the chamber walls. The photoelectrons will again strike the anode to produce more UV,

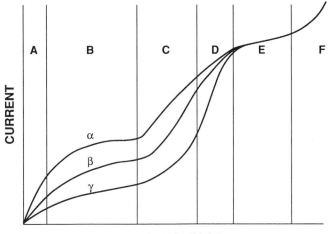

Fig. 7.2. A composite curve illustrating the current output as a result of increasing voltages for different radiations. (A) Region of recombination, (B) region of saturation, (C) proportional region, (D) region of limited proportionality, (E) Geiger region, and (F) continuous discharge.

and hence an avalanche spreads along the entire length of the anode. The amplification factor can be as high as 10^{10}. During the avalanche, however, the lightweight electrons are quickly attracted to the anode, whereas a sheath of slow-moving heavy positive ions builds up around the anode. As a result, the voltage gradient falls below the value necessary for ion multiplication, and therefore the avalanche is terminated. All this occurs in less than 0.5 microsecond, and the counter is left insensitive and must recover before another event can be counted.

Recovery begins with the migration of the positive ions toward the cathode (i.e., chamber wall) and takes about 200 microseconds at a gas pressure of 0.1 atmosphere, which is equal to the dead time of the counter. As the positive ions approach the cathode, secondary electrons may be emitted from the surface of the cathode, which then set another discharge just about 200 microseconds after the previous one. Such repetitive discharges that are due to secondary electrons are independent of the sources of radiation that the counter is intended to measure. The emission of secondary electrons is suppressed by a technique known as *quenching* to eliminate repetitive counter discharges (see later).

As the applied voltage is increased beyond the Geiger region, a single ionizing event produces a series of repetitive discharges leading to what is called *spontaneous discharge*. This region is called the *region of continuous discharge* (see Fig. 7.2). Operation of a detector in this region may cause damage to the detector.

Ionization Chambers

Ionization chambers are operated at voltages in the saturation region that spans 50–300 V. The detector is a cylindrical chamber filled with air or a gas, sometimes at high pressure. A central wire and the chamber act as the electrodes. Ionization chambers are primarily used for measuring high-intensity radiation such as x-ray beams and high activity of radiopharmaceuticals. Cutie Pie meters, dose calibrators, and pocket dosimeters are the common ionization chambers used in nuclear medicine.

Cutie Pie Survey Meter

The Cutie Pie survey meter is made of an outer metallic cylindrical electrode and a central wire. It uses air for ionization and is operated with a battery. It is used to measure the radiation exposure in the range of mR/hr to R/hr. It is primarily used to monitor the exposure at high radiation levels such as those from x-ray beams and ^{99}Mo–^{99m}Tc generators.

Dose Calibrator

The dose calibrator is one of the most essential instruments in nuclear medicine for measuring the activity of radionuclides and radiopharmaceuticals. It is a cylindrically shaped, sealed chamber with a central well and is filled

Fig. 7.3. A typical dose calibrator. (Courtesy of Capintec, Inc., Ramsey, NJ.)

with argon and traces of halogen at high pressure ($\sim$25–30 atmospheres). Its operating voltage is about 150 V. A typical dose calibrator is shown in Figure 7.3.

Because radiations of different types and energies produce different amounts of ionization (hence current), equal activities of different radionuclides generate different quantities of current. For example, the magnitude of the current produced by 1 mCi (37 MBq) of ^{99m}Tc differs from that produced by 1 mCi (37 MBq) of ^{131}I. Isotope selectors provided on the dose calibrator are the feedback resistors to compensate for the differences in ionization (current) produced by different radionuclides so that equal activities produce the same reading. In most dose calibrators, isotope selectors for common radionuclides are push-button type, whereas those for other radionuclides are set by a continuous dial. An activity range selector is a variable resistor that adjusts the range of activity (μCi, mCi, Ci, MBq, GBq, kBq) for display.

Dose calibrators must be checked for constancy, accuracy, linearity of their operation, and geometry of the sample. These tests are required by the Nuclear Regulatory Commission (NRC) in the United States. The frequency of tests of the dose calibrator is as follows:

1. Constancy (daily)
2. Accuracy (at installation, annually, and after repairs)
3. Linearity (at installation, quarterly, and after repairs)
4. Geometry (at installation and after repairs)

Daily constancy check is performed by measuring a long-lived radioactivity (e.g., ^{137}Cs) in the dose calibrator and observing the variation not to exceed $\pm 10\%$ relative to the previous day reading. If the variation exceeds $\pm 10\%$, the unit must be repaired or replaced.

Accuracy of the dose calibrator is determined by measuring the activity of at least two long-lived radionuclides (e.g., ^{137}Cs and ^{57}Co) certified by the National Institute of Standards and Technology (NIST) in the dose calibrator and comparing the measured activity with the activity reported by the NIST. The measured value should not differ from the theoretical value by more than $\pm 10\%$. If it exceeds $\pm 10\%$, the unit must be repaired or replaced.

The linearity test indicates the dose calibrator's ability to measure the activity accurately over a wide range of values. It is carried out by assaying a very high-activity ^{99m}Tc source (covering the range of activities routinely used for radiopharmaceuticals) at 0, 6, 24, 30, and 48 hr, and plotting the activity against time. The measured activity should agree with the calculated activity of ^{99m}Tc indicating the linearity of response in the range of activities of interest. Normally, dose calibrators are linear in response for activities up to 200 mCi (7.4 GBq) to 2 Ci (74 GBq), depending on the chamber geometry and electronics of the dose calibrator and tend to underestimate at higher activities. In the case of nonlinearity of response, correction factors must be determined and applied to the measured activities, if the error exceeds $\pm 10\%$.

Variations in sample volumes or in geometric configurations of the container can affect the accuracy of measurements in a dose calibrator, particularly for low-energy radiations. Thus, 1 mCi (37 MBq) in 1-ml or 30-ml volume, or 1 mCi (27 MBq) in 1-cc syringe, 10-cc syringe or 10-cc vial, or in containers of different materials (plastic or glass) may give different readings in the dose calibrator. Correction factors must be determined for these geometric variations and applied to the measured activities, if the error exceeds $\pm 10\%$.

Pocket Dosimeter

The pocket dosimeter operates on the principle of a charged electroscope equipped with a scale inside. It consists of a quartz fiber electroscope inside the chamber. Initially, the dosimeter is fully charged by means of an external power supply, and the scale then reads zero. After exposure to radiation, charge is lost, and the loss of charge is proportional to the amount of radiation exposure, which is read on the inside scale in mR/hr. This reading can be seen through a viewing window at the end of the dosimeter. After complete discharge of the dosimeter, it can be charged and used again. It is primarily used to determine personnel exposure while working with radiation and has the advantage of giving immediate readings. These dosimeters are available in full-scale readings of 200 mR, 500 mR, and 1 R.

Geiger–Müller Counters

The Geiger–Müller (GM) counter operates in the Geiger region of the voltage, as shown in Figure 7.2. As already mentioned, in this region, an avalanche of ionizations occurs as a result of high voltage. Once an ionization is initiated, the avalanche of ionizations can lead to repetitive discharges unless the process is interrupted by the quenching technique. An electronic technique of quenching can be applied in which the voltage applied to the GM tube is temporarily reduced below the Geiger region until all ion pairs return to their de-excited states. This happens in a few tenths of a millisecond. The original voltage is then restored for the detection of the next event. This technique is no longer in use.

The common technique of quenching is to add a small quantity of a quenching gas to the argon counting gas. Either organic solvent vapors (e.g., ethyl alcohol, xylene, or isobutane) or halogen gases (chlorine or bromine) are commonly used as the quenching gas. These molecules transfer electrons to the "positive" ion cloud and become themselves ionized. Ionized molecules of the quenching gas migrate to and dislodge electrons from the cathode. When these electrons neutralize the ionized molecules of the quenching gas, energy is released, which causes the dissociation of the molecules of the gas but with no UV emissions to prolong the avalanche. This prevents the continuous

Fig. 7.4. An end-window–type Geiger–Müller counter. (Courtesy of Nuclear Associates, Carle Place, NY.)

discharge of the GM counter. Organic molecules are more effective quenchers but dissociate irreversibly and therefore give a limited lifetime for the GM tube ($\sim 10^8 - 10^{10}$ pulses). In contrast, dissociated inorganic molecules recombine to form the original molecules, and therefore halogen-quenched GM tubes have infinite useful lifetime.

The probes of GM counters can be either end-window type or side-window type. An end-window–type GM counter is shown in Figure 7.4. The window is made of thin mica, and gases such as argon, methane, helium, and neon are commonly used as the counting gas. Different shapes of GM probes are available, such as cylindrical and pancake types. Some GM probes are provided with a metal cover that stops all β particles and low-energy γ-radiations so that only high-energy photons are detected. Without the cover, β-particles and low-energy γ-rays can be detected. The GM counter is usually battery operated at a voltage of 1000–1200 V. The meter connected to the GM probe gives readings in mR/hr or counts per minute. Some counters are equipped with audible alarms or flashing light alarms that are triggered by radiation above a preset intensity. The latter kind is often used to monitor the radiation level in different work areas and is called an *area monitor.*

The GM survey meters are more sensitive than ionization chambers, but they cannot discriminate between energies. These counters are almost 100% efficient for counting α and β particles but have only 1% to 2% efficiency for counting γ- and x-rays. The dead time, or resolving time (Chapter 8), of the GM counters is about 100 to 500 microseconds. This limits the count rates to about 15,000–20,000 counts per minute (cpm) for these counters, and at higher activities they tend to saturate, thus losing counts. The GM counters are normally used for area survey for contamination with low-level activity. According to the NRC regulations, these survey meters must be calibrated annually with standard calibrated sources such as ^{137}Cs.

Questions

1. Describe the principles of gas-filled detectors.
2. What are the differences between an ionization chamber and a Geiger–Müller counter?
3. What is the function of a push-button isotope selector on a dose calibrator?
4. Can you discriminate between 140-keV γ-rays, 364-keV γ-rays, and 5-MeV α particles using a GM counter?
5. What type of instruments would you use for:
   (a) Survey of the laboratory?
   (b) X-ray beam exposure?
   (c) Area survey around x-ray room?
   (d) Spill of 50 μCi (1.85 MBq) of ^{201}Tl?
   (e) Background radiation?
   (f) Radiation survey of a diagnostic x-ray installation?
6. (a) Why are halogen gases added to GM counters?
   (b) What is the typical dead time for GM counters?
   (c) How often do the GM counters need to be calibrated?
   (d) Why cannot the GM counters be used for detecting high-activity samples?
7. What are the typical voltages applied to the ionization chambers and GM counters?
8. Describe the various tests of the dose calibrator and mention the frequency of each test as required by the NRC.

Suggested Readings

Hendee WR. *Medical Radiation Physics*. 2nd ed. Chicago: Year Book Medical Publishers, Inc; 1979.

Ouseph PJ. *Introduction to Nuclear Radiation Detectors*. New York: Plenum Press; 1975.

Robinson CV. Geiger–Müller and proportional counters. In: Hine GJ, ed. *Instrumentation in Nuclear Medicine*. New York: Academic Press, 1967; 57–72.

Rollo FD, ed. *Nuclear Medicine Physics, Instrumentation and Agents*. St Louis: Mosby; 1977.

Sorensen JA, Phelps ME. *Physics in Nuclear Medicine*. 2nd ed. New York: Grune & Stratton; 1987.

CHAPTER 8

Scintillation and Semiconductor Counters

Principles of Scintillation Detectors

As stated in Chapter 7, the detection efficiency of γ- and x-rays in gas detectors is very low, because these penetrating radiations travel through the low-density gas with little interaction. To improve counting efficiency for these radiations, solid and liquid scintillation detectors with high density are used. These detectors have the unique property of emitting scintillations or flashes of light after absorbing γ- or x-radiations. The light photons produced are converted to an electrical pulse by means of a photomultiplier (PM) tube (described later). The pulse is then amplified by a linear amplifier, sorted by a pulse-height analyzer (PHA) and then registered as a count. Different solid or liquid detectors are used for different types of radiation. For example, sodium iodide crystals containing a trace of thallium (NaI[Tl]) are used for γ- and x-ray detection, whereas organic crystals such as anthracene and plastic fluors are used for β-particle detection.

In liquid scintillation counting, a β^--emitting radioactive sample and an organic scintillator are dissolved in a solvent. The β^- particle interacts with solvent molecules emitting electrons. The latter interact with the organic scintillator, whereby light photons are emitted, which are then directed to two PM tubes coupled in coincidence. A pulse is produced by the PM tube, which is registered as a count, as in solid scintillation counting.

Organic scintillators usually have a lower density and, hence, a lower counting efficiency than inorganic scintillators. However, the decay time for light emission for organic scintillators is much shorter than that for inorganic scintillators. For example, the decay time for NaI(Tl) is 0.25 μsec and that for anthracene is 0.026 μsec. The faster decay time permits the use of organic scintillators at higher count rates.

Solid Scintillation Counters

A basic solid scintillation counter consists of a NaI(Tl) detector, a PM tube, a preamplifier, a linear amplifier, a PHA, and a recording device (Fig. 8.1). Each of these components is described in detail next.

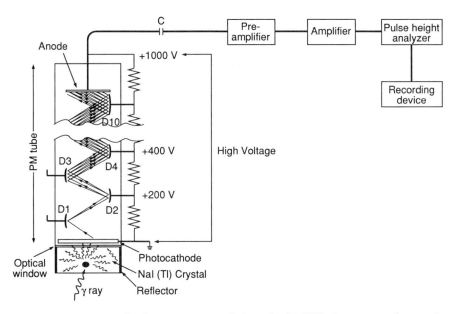

Fig. 8.1. A basic scintillation counter consisting of a NaI(Tl) detector, a photomultiplier (PM) tube, a preamplifier, a linear amplifier, a pulse-height analyzer, and a recording device. The high voltage applied to the PM tube is typically 1000 V.

NaI(Tl) Detector

Pure sodium iodide does not produce any scintillation after interaction with γ-radiations at room temperature. However, if it is doped with a trace amount (0.1%–0.4%) of thallium as an activator, NaI(Tl) becomes quite efficient in producing light photons after γ-radiations interact with it. γ-rays or x-rays interact with NaI(Tl) via photoelectric, Compton, and/or pair production mechanisms, whereby the NaI molecules are raised to higher energy states through ionization or excitation. The excited states return to ground states by emitting light photons. Approximately 20–30 light photons are produced per 1 keV of energy.

The choice of NaI(Tl) crystals for γ-ray detection is primarily due to the high density (3.67 g/cm^3) of the crystal and the high atomic number of iodine ($Z = 53$), compared to organic scintillators. However, NaI(Tl) crystals are hygroscopic, and absorbed water causes color changes that reduce light transmission to the PM tubes. Therefore, the NaI(Tl) crystals are hermetically sealed in aluminum containers. Also, the entrance and side of the crystals are coated with a reflective substance (e.g., magnesium oxide) so that light photons are reflected toward the photocathode of the PM tube (see later). These crystals are fragile, and must be handled with care. Room temperature should not be changed abruptly, because such changes in temperature can cause cracks in the crystal.

The NaI(Tl) detectors are made of various sizes for different types of instrument. They vary from 3.8 to 50 cm in diameter and 0.63 to 23 cm in thickness. In thyroid probes, well counters and rectilinear scanners, smaller and thicker crystals are used, whereas larger and thinner crystals are employed in scintillation or gamma cameras.

Photomultiplier Tube

A PM tube consists of a light-sensitive photocathode at one end, a series (usually 10) of metallic electrodes known as *dynodes* in the middle, and an anode at the other end—all enclosed in a vacuum glass tube (see Fig. 8.1). The photocathode is usually an alloy of cesium and antimony that releases electrons after absorption of light photons. The PM tube is fixed on to the NaI(Tl) crystal with the photocathode facing the crystal by a special optical grease or connected to the crystal using light pipes.

A high voltage of ~ 1000 V is applied between the photocathode and the anode of the PM tube in steps of 50–150 V between dynodes (see Fig. 8.1). When light photons from the NaI(Tl) crystal strike the photocathode, photoelectrons are emitted, which are accelerated toward the next closest (i.e., first) dynode by the voltage difference between the electrodes. Approximately 1 to 3 photoelectrons are produced from the photocathode per 7 to 10 light photons. Each of these photoelectrons is accelerated to the second dynode and emits 2 to 4 electrons upon impingement. The accelerated electrons strike the successive dynodes, and more electrons are emitted. The process of multiplication continues until the last dynode is reached, where a pulse of 10^5 to 10^8 electrons is produced. The pulse is then attracted to the anode and finally delivered to the preamplifier. The amplitude of the pulse is proportional to the number of light photons received by the photocathode and in turn to the energy of the γ-ray photon absorbed in the crystal. The applied voltage must be very stable, because slight changes in dynode voltage cause a great variation in electron multiplication factor.

Preamplifier

The pulse from the PM tube is small in amplitude and is initially amplified by a preamplifier. The preamplifier adjusts the voltage of the pulse (pulse shaping) and matches impedance level between the detector and the subsequent circuits so that the pulse is appropriately processed by the system.

Linear Amplifier

A linear amplifier amplifies further the signal from the preamplifier and delivers it to the pulse height analyzer for analysis of its amplitude. The amplification of the pulse is given by the amplifier gain expressed as the ratio of the amplitude of the outgoing pulse to that of the initial pulse from the PM tube. The amplifier gains are given in the range of 1 to 1000 by gain control knobs provided on the amplifier. The output pulses normally have amplitudes of up to 10 V.

Pulse-Height Analyzer

γ-rays of different energies can arise from a source of the same or different radionuclides or can be due to scattering of γ-rays in the source and the detector. Thus, the pulses coming out of the amplifier may differ in magnitude. A PHA is a device that selects for counting only those pulses falling within preselected voltage intervals or "channels" and rejects all others (see Fig. 8.1). Pulses corresponding to γ-ray energies of interest are selected by energy discriminator knobs, known as the *lower level* or *upper level*, or the *baseline* and *window*, provided on the PHA, and are ultimately delivered to the recording devices such as scalers, computers, films, and so on.

There are two modes of counting using PHAs: differential and integral. In differential counting, only pulses of preselected energies are counted by appropriate selection of lower and upper level knobs (discriminators) or the baseline and window. In scintillation cameras, however, differential counting is achieved by a peak voltage knob and a percent window knob. The peak voltage knob sets the energy of the desired γ-ray, and the percent window knob sets the window width in percentage of the γ-ray energy, which is placed symmetrically on each side of the peak voltage.

In integral counting, γ-rays of all energies or all γ-rays of energies above a certain energy are counted by setting the appropriate lower level or baseline and bypassing the upper level or window mechanism.

A PHA normally selects only one range of pulses corresponding to only one γ-ray energy by means of differential counting. Such a PHA is called a single-channel analyzer (SCA). A multichannel analyzer (MCA) is a device that can simultaneously sort pulses into many predetermined voltage ranges or channels. By using an MCA, one obtains a simultaneous spectrum of different γ-ray energies from a radioactive source.

Display or Storage

Pulses processed by the PHA can be displayed on a cathode ray tube (CRT) or can be counted for a preset count or time by a scaler-timer device. A rate meter can be used to display the pulses in terms of counts per minute (cpm) or counts per second (cps). In scintillation cameras, pulses are used to form the image on a CRT and polaroid or x-ray films. These pulses can also be stored in a computer or on a magnetic tape or laser disc for processing later.

Gamma Ray Spectrometry

Pulses are generated by the PM tube and associated electronics after the γ-ray energy is absorbed in the NaI(Tl) detector. Because γ-rays interact with the NaI(Tl) detector by photoelectric, Compton, and pair production mechanisms, and also because various scattered radiations from outside the detec-

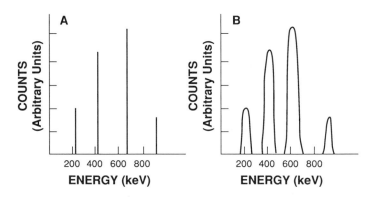

Fig. 8.2. γ-ray spectra. (A) An ideal spectrum would represent the different γ-rays as lines. (B) An actual spectrum showing the spread of the photopeak that is due to statistical fluctuations in the pulse formation.

tor may interact with the detector, a distribution of pulse heights will be obtained depicting a spectrum of γ-ray energies. Such a γ-ray spectrum may result from a single γ-ray or from many γ-rays in a sample. Different features of this spectrum are discussed later.

Photopeak

In an ideal situation, if the γ-ray photon energy is absorbed by the photoelectric mechanism and each γ-ray photon yields a pulse of the same height, then each γ-ray would be seen as a line on the γ-ray spectrum (Fig. 8.2A). In reality, the photopeak is broader, which is due to various statistical variations in the process of forming the pulses. These random fluctuations arise from the following conditions:

1. Because 20–30 light photons are produced for every keV of γ-ray energy absorbed, there is a statistical variation in the number of light photons produced by the absorption of a given γ-ray energy. Also, statistically all light photons produced in the detector may not strike the photocathode.
2. As already stated, 7 to 10 light photons are required to release 1 to 3 photoelectrons from the photocathode. Therefore, the number of photoelectrons that one γ-ray will produce may vary from one event to another.
3. The number of electrons released from the successive dynodes by impingement of each electron from the previous dynode varies from two to four, and therefore pulse heights from the PM tube will vary from one γ-ray to the next of the same energy.

All of the preceding statistical fluctuations in generating a pulse cause a spread in the photopeak (see Fig. 8.2B). A typical spectrum of the 662-keV γ-ray of ^{137}Cs is shown in Figure 8.3.

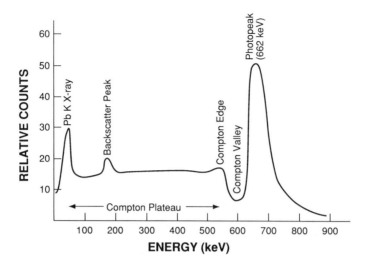

Fig. 8.3. A typical spectrum of the 662-keV γ-ray of ^{137}Cs illustrating the photopeak, Compton plateau, Compton edge, Compton valley, backscatter, and characteristic lead K x-ray peaks.

Compton Valley, Edge, and Plateau

When γ-rays interact with the NaI(Tl) detector via Compton scattering and scattered photons escape from the detector, the Compton electrons result in a pulse height that is smaller than that of the photopeak. The Compton electrons, however, can have variable energies from zero to E_{max}, where E_{max} is the kinetic energy of those electrons that are produced by the 180° Compton backscattering of the γ-ray photon in the detector. At relatively high photon energy, E_{max} is given by the photon energy minus 256 keV (Equation 6.2). Thus, the γ-ray spectrum will show a continuum of pulses corresponding to Compton electron energies between zero and E_{max}. The peak at E_{max} is called the *Compton edge*, and the portion of the spectrum below the Compton edge up to about zero energy is called the *Compton plateau* (see Fig. 8.3). The portion of the spectrum between the photopeak and the Compton edge is called the *Compton valley*, which results from multiple Compton scattering of the same γ-ray in the detector yielding pulses of amplitudes in this region.

The relative heights of the photopeak and the Compton edge depend on the photon energy as well as the size of the NaI(Tl) detector. At low energies, photoelectric effect predominates over Compton scattering, whereas at higher energies, the latter becomes predominant. In larger detectors, γ-rays may undergo multiple Compton scattering, which can add up to the absorption of the total photon energy identical to the photoelectric effect. This increases the contribution to the photopeak and decreases to the Compton plateau.

Characteristic X-Ray Peak

Photoelectric interactions of the γ-ray photons in the lead shield around the detector can lead to the ejection of the K-shell electrons, followed by transition of electrons from the upper shells, mainly the L shell, to the K shell. The difference in binding energy between the K-shell electron (~ 88 keV) and the L-shell electron (~ 16 keV) appears as lead K x-ray of ~ 72 keV. These characteristic x-ray photons may be directed toward the crystal and absorbed in it and may appear as a peak in the γ-ray spectrum (see Fig. 8.3). These photons can be reduced by increasing the distance between the detector and the shielding material.

Backscatter Peak

When γ-ray photons, before striking the detector, are scattered at 180° by Compton scattering in lead shielding and housing, and the scattered photons are absorbed in the detector, then a peak, called the *backscatter peak*, appears in the γ-ray spectrum (see Fig. 8.3). This peak can be mostly eliminated by increasing the distance between the shield and the detector.

Iodine Escape Peak

Photoelectric interaction of γ-ray photons with iodine atoms of the NaI(Tl) detector usually results in the emission of characteristic K x-rays. These x-ray photons may escape the detector, resulting in a peak equivalent to photon energy minus 28 keV (binding energy of the K-shell electron of iodine). This is called the *iodine escape peak*, which appears about 28 keV below the photopeak (Fig. 8.4). This peak becomes prominent when the energy of the photon is less than about 200 keV, because, at energies above 200 keV, the iodine

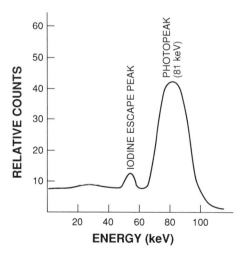

Fig. 8.4. A spectrum of 81-keV γ-ray of ^{133}Xe showing an iodine escape peak.

escape peak would fall within the width of the photopeak, which is due to small differences between the two peaks.

Annihilation Peak

γ-rays with energy greater than 1.02 MeV may undergo pair production in the detector in which a β^+- and a β^- particle are produced. The β^+ particles are annihilated to produce two 511-keV photons, which appear as photopeaks in the γ-ray spectrum. If, however, one of the 511-keV photons escapes from the detector, then a peak, called the *single-escape peak*, corresponding to the primary photon energy minus 511 keV, will appear in the spectrum. If both annihilation photons escape, then a *double-escape peak* results, corresponding to the primary photon energy minus 1.02 MeV. Larger detectors can prevent the escape of the annihilation radiations.

Coincidence Peak

A *coincidence* or *sum peak* results when more than one photon is absorbed simultaneously in the detector to be considered as a single event. The peak equals the sum of the energies of the photons. Such situations occur with radionuclides that have short-lived isomeric states and thus emit γ-rays in cascade. For example, ^{111}In emits 171- and 245-keV photons, which can result in a sum peak of 416 keV (Fig. 8.5). These peaks can be reduced by counting the samples at larger distances between the source and the detector or by using smaller detectors so that the likelihood of two photons striking the detector at the same time is reduced.

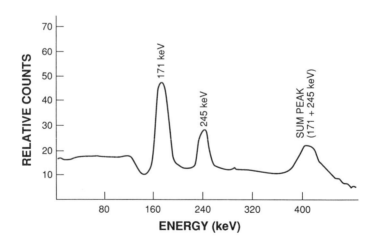

Fig. 8.5. A spectrum of ^{111}In with 171- and 245-keV photons showing a coincidence (sum) peak at 416 keV.

Liquid Scintillation Counters

Low-energy β^- particles are normally absorbed within the source and in the window and walls of the detectors, and therefore β^- emitters are difficult to detect in gas or solid detectors. For this reason, β^--emitting radionuclides are counted using the liquid scintillation technique in which the radioactive sample is mixed with a scintillating material. A sample vial containing the liquid scintillator and the radioactive sample of interest is placed between two PM tubes connected in coincidence (Fig. 8.6). Each PM tube receives the light photons emitted by the interaction of the β^- particle with the scintillator and converts them into a pulse, which is further amplified by an amplifier. The two amplified pulses are then delivered to the coincidence circuit that contains a PHA to analyze the pulse height for acceptance. A count is registered in the scaler if an event is recorded in both PM tubes simultaneously. Such coincidence counting reduces the background counts that are due to noise.

The liquid scintillation solution is prepared by dissolving a primary scintillating solute or fluor and often a secondary fluor in a solvent. The radioactive sample is added to and thoroughly mixed with the scintillating solution for counting. The primary fluors include 2,5-diphenyloxazole (PPO), 2,5-*bis*-2-(5-T-butyl-benzoxazolyl)-thiophene (BBOT), and *p*-terphenyl, of which PPO is most commonly used in a concentration of 5 g/L.

Toluene, xylene, and dioxane are the most common solvents that easily dissolve the primary fluor and often the radioactive sample, which is a requirement for a good solvent. These solvents, however, are poorly miscible in water, and therefore their disposal in the sewer system is prohibited. Nowadays, the biodegradable solvents such as linear alkylbenzene and phenylxyly-

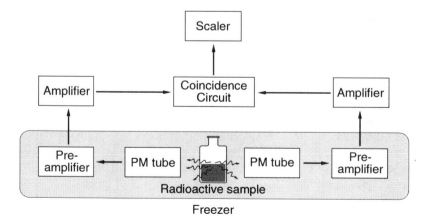

Fig. 8.6. A schematic diagram of a liquid scintillation counting system. Light photons emitted from the sample strike the two photomultiplier tubes to produce pulses. Only coincident pulses are counted.

lethane are widely used. Vials are usually glass or plastic, but the latter is not used when toluene or xylene is used as a solvent because the solvent tends to dissolve plastic.

When radiations pass through the solvent, electrons are released from the solvent molecules after absorption of radiation energies. These electrons transfer energy to primary fluor molecules, which then emit light photons for further processing by PM tubes and associated electronics. The wavelength of these light photons is somewhat shorter than required for the spectral sensitivity of the photocathode of the PM tube. This mismatch is rectified by adding a secondary solute, called the *wavelength shifter*, to the scintillating solution. The wavelength shifter absorbs the light photons emitted by the primary fluor and re-emits them with a longer wavelength, which is more suitable for the photocathode of the PM tube. The compound 1,4-*bis*-2-(5-phenyloxazolyl)-benzene (POPOP) is most commonly used as a secondary solute in a concentration of about 0.1%.

Attempts are always made to keep the radioactive sample in solution with the liquid scintillator. Solubilizing agents are added to improve dissolution of specific samples, and the common example is the hydroxide of Hyamine 10-X used in counting tissue samples.

In liquid scintillation counting, quenching is a problem caused by interference with the production and transmission of light, which ultimately reduces the detection efficiency of the system. Quenching can be of the following types:

1. *Chemical type*, resulting from interference in energy transfer by substances such as samples or extraneous materials (e.g., dissolved O_2)
2. *Color type*, resulting from absorption of light photons by colored substances, such as hemoglobin, before striking the PM tube
3. *Dilution type*, resulting from relatively large dilution of the scintillation mixture, in which case light photons may be absorbed by the solvent, such as water
4. *Optical type*, resulting from absorption of light by a dirty vial containing frost or fingerprints.

Quenching must be corrected for obtaining accurate counting of samples, and three methods have been adopted for this purpose, namely, internal standard method, channel ratio method, and external standard method. The readers are referred to reference physics books for details of these methods.

The liquid scintillation counting systems are provided with automatic sample changers for counting as many as 500 samples. Also, one to five PHAs are available on a liquid scintillation counter, so that β^- particles of different energies can be counted simultaneously by using different baselines and windows on each PHA. The β^--emitters, ^{3}H, ^{14}C, ^{32}P, and ^{35}S, are commonly detected by liquid scintillation counting. Whereas the counting efficiencies of ^{3}H ($E_{max} = 0.018$ MeV) and ^{32}P ($E_{max} = 1.71$ MeV) are ~ 60–70% and $\sim 100\%$ respectively, it is negligible for γ- and x-rays.

Semiconductor Detectors

Semiconductor detectors or solid-state detectors are made of germanium or silicon elements commonly doped with lithium. These detectors are designated as Ge(Li) or Si(Li) detectors, of which the former is commonly used for γ-ray detection and the latter for α-particle detection. The basic principle of operation of these detectors involves ionization of the semiconductor atoms, as in gas detectors. Ionizations produced in the detector by radiation are collected as current and converted to voltage pulses through a resistor by the application of a voltage. The pulses are then amplified and counted. The size of the pulse is proportional to the radiation energy absorbed in the detector.

Because semiconductors are much denser than gases, they are more efficient for x- and γ-ray detection than gas detectors. Also, in semiconductor detectors, each ionization requires only about 3 eV compared to 35 eV in gas detectors. Thus, almost 10 times more ions are produced in semiconductor detectors than in gas detectors for a given γ-ray energy, thus yielding a better energy resolution of γ-ray photons of closer energies.

Fabrication of Ge(Li) and Si(Li) detectors is quite time-consuming and expensive. Detectors are small in size (e.g., about 4 cm in diameter by 4 cm in thickness for Ge(Li) detectors), and therefore, the detection efficiency for x- and γ-rays is much lower than that of NaI(Tl) detectors.

Thermal noise at room temperature introduces a high background, which can obscure the sample counts but is reduced at low temperature. Therefore, these detectors are operated at low temperature, usually employing liquid nitrogen ($-196°C$ or $77°K$). A disadvantage of these detectors is that liquid nitrogen evaporates and must be replenished periodically—typically weekly.

Semiconductor detectors are most useful in differentiating among photon energies because of the high-energy resolution, particularly in detecting radionuclidic contamination. These detectors are not in common use in nuclear medicine.

Cadmium-Telluride Detectors

Cadmium-Telluride (CdTe) detectors are made of Cd and Te metals and provide very high efficiency for γ-ray detection because of their high atomic numbers. For reasons of high detection efficiency, these detectors can be made as small as 2-mm thick and 2-mm diameter with almost 100% efficiency for 100-keV photons. The energy resolution for these detectors is very good for a wide range of γ-ray energies. These detectors can be operated at room temperature. The electronics used is similar to those of semiconductor detectors. Different types of handheld probes have been devised for various purposes. One probe, called the Neoprobe 1000, is used for the detection of metastatic sites containing radioactivity (e.g., ^{125}I-labeled monoclonal antibody) during surgery for removal by incision.

Characteristics of Counting Systems

Detection of radiation and therefore counting of radioactive samples is affected by different characteristics of the detector and the associated electronics. The following is a discussion of these properties.

Energy Resolution

As already mentioned, even though γ-rays of the same energy are absorbed in the NaI(Tl) detector by the photoelectric effect, pulses of different amplitudes are produced because of the statistical variations in the production of light photons in the detector and photoelectrons and electrons in the PM tube. This results in the broadening of the photopeak. The width of the peak or the sharpness of the peak predicts the ability of the NaI(Tl) spectrometer to discriminate the γ-ray photons of dissimilar energies. A similar situation exists for semiconductor detectors where the number of ionizations may vary from one γ-ray to another of the same energy, leading to the broadening of the peak.

The energy resolution of a system is given by the full width at half-maximum (FWHM) amplitude of the photopeak and is expressed as a percentage of the photon energy as follows:

$$\text{Energy resolution (\%)} = \frac{\text{FWHM}}{E_{\gamma}} \times 100 \tag{8.1}$$

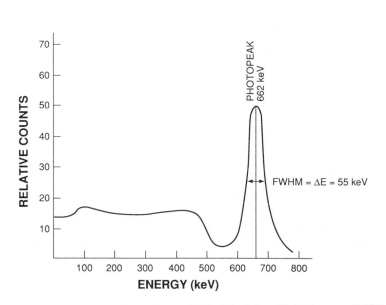

Fig. 8.7. The full width at half maximum (FWHM) of the 662-keV γ-ray of ^{137}Cs in a NaI(Tl) detector.

where E_γ is the energy of the γ-ray photon. In Figure 8.7, FWHM is 55 keV for the 662-keV peak of ^{137}Cs; therefore,

$$\text{Energy resolution (\%)} = \frac{55}{662} \times 100$$

$$= 8.3\%$$

The energy resolution depends on the photon energy and the size of the detector. The higher the photon energy, the better the energy resolution (i.e., smaller FWHM), because of the decrease in the percentage of statistical variations in the pulse production. The FWHM (%) in NaI(Tl) detectors is about 7% to 10% for the 662-keV γ-ray of ^{137}Cs and 10% to 14% for the 140-keV γ-ray of ^{99m}Tc. In contrast, the FWHM (%) in Ge(Li) detectors is about 0.42% for 140-keV γ-rays and about 0.2% for photons of more than 1 MeV.

Detection Efficiency

The detection efficiency of a counter is given by the observed count rate divided by the disintegration rate of a radioactive sample. The count rate of a sample differs from the disintegration rate because of several factors. Radiations from a source are emitted isotropically around 360°, but only a fraction of all photons emitted strikes the detector, depending on the solid angle subtended by the detector on the source. Only a fraction of all photons striking the detector may interact in the detector and produce pulses. Only a fraction of all pulses produces a single photopeak. Furthermore, the count rate is affected by the abundance of a particular radiation from a radionuclide. Considering these factors, the overall counting efficiency of a counter for a radiation is given by the following expression:

$$\text{Efficiency} = f_i \times f_p \times f_g \times N_i \tag{8.2}$$

where f_i is the intrinsic efficiency; f_p is the photopeak efficiency, or photofraction; f_g is the geometric efficiency; and N_i is the abundance of the radiation in question. N_i is available in literature on Tables of Isotopes.

Intrinsic Efficiency

The fraction of all radiations of a given type and energy impinging on the detector that interacts with it to produce pulses is called the *intrinsic efficiency*, f_i, of the detector:

$$f_i = \frac{\text{No. of radiations detected by the detector}}{\text{No. of radiations impinging on the detector}}$$

$$= \frac{\text{All counts under the entire spectrum}}{\text{No. of radiations impinging on the detector}} \tag{8.3}$$

Intrinsic efficiency depends on the type and energy of the radiation and the linear attenuation coefficient (μ) and thickness of the detector. The value of f_i is almost 1 for low-energy γ-rays and thicker detectors. The f_i tends to 0 for high-energy γ-rays and thinner detectors. These conditions apply to both NaI(Tl) and semiconductor detectors. The intrinsic efficiency of gas detectors is almost unity for α and β particles but is about 0.01 (1%) for γ- and x-rays.

Photopeak Efficiency or Photofraction

The fraction of all detected γ-rays that contributes only to the photopeak is called the *photopeak efficiency*, or *photofraction* (f_p). It is given as the total photopeak area divided by the total area under the entire spectrum:

$$f_p = \frac{\text{All counts under the photopeak}}{\text{All photons detected by the detector}} \tag{8.4}$$

This value is affected by all factors that influence photoelectric effect, such as the size and composition of the detector and γ-ray energy, but is primarily determined by the discriminator settings on the PHA. The f_p increases with increasing window settings.

Geometric Efficiency

Radiations from a source are emitted uniformly with equal intensity in all directions. If a source of radiation is placed at a distance from a detector, then only a fraction of all radiations emitted from the source will be detected by the detector. This fraction is determined by the solid angle subtended by the detector on the source. The geometric efficiency, f_g, is equal to the number of radiations striking the detector divided by the total number of radiations emitted by the source. Thus,

$$f_g = \frac{\text{No. of radiations striking the detector}}{\text{Total number of radiations emitted by the source}}$$

For a circular detector with radius r, it is equal to the area πr^2 of the detector divided by the total spherical area $4\pi R^2$, where R is the distance between a point source S and the detector D (Fig. 8.8).

$$f_g = \frac{\pi r^2}{4\pi R^2} \tag{8.6}$$

As the distance R between the source and the detector increases, the f_g decreases, according to the inverse square law; that is, $f_g \propto 1/R^2$ (Fig. 8.8A). Thus the f_g at $2R$ is one fourth of the f_g at R. The value of f_g increases with the size of the detector. Also, the finite size of the radiation source affects the f_g values.

When the source and the detector are in close contact, the f_g tends to be about 50% (Fig. 8.8B). In the case of gamma well counters and liquid scintillation counters, the f_g approaches 100% (Fig. 8.8C).

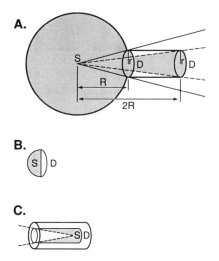

Fig. 8.8. Illustration of geometric efficiency, f_g, of a detector D with a circular area, πr^2, where r is the radius of the detector. (A) The detector D placed at a distance R from the point source S has an f_g four times greater than the f_g when the detector is placed at a distance $2R$. (B) When the source and the detector are in close contact, the f_g is about 50%. (C) When the source is well inside the detector as in a well counter, the f_g approaches 100%.

Dead Time

Each counting system takes a certain amount of time to process a radiation event, starting from interaction of radiation with the detector all the way up to forming a pulse and ultimately recording it. The counter remains insensitive to a second event for this period of time, that is, if a second radiation arrives during this time, the counter cannot process it. This period is called the *dead time*. When the counter recovers after this period, only then can a second radiation be detected. During the dead time, *pulse pileup* can occur when pulses from two subsequent γ-rays combine to form a pulse of amplitude that is beyond the PHA window setting and thus is rejected. Radiations lost by pulse pileup during the dead time are called the *dead-time losses*, or *coincidence losses*.

The dead time of a counting system may arise from different components of the entire counting system: detector, PHA, PM tube, scaler, computer interface, and so on. While Geiger–Müller (GM) detectors have a longer dead time of 100 to 500 microseconds, the typical values for NaI(Tl) and semiconductors are of the order of 0.5 to 5 microseconds and for liquid scintillators, ~0.1 to 1 microsecond (Sorensen and Phelps, 1987).

Based on how successive pulses are processed owing to the dead time, the counting systems fall into two categories: *paralyzable* and *nonparalyzable* (Sorensen and Phelps, 1987). In paralyzable systems, each event sets its own

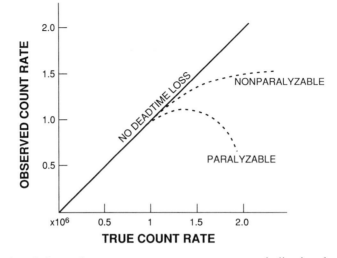

Fig. 8.9. Plot of observed count rates versus true count rates indicating the dead time loss in paralyzable and nonparalyzable systems.

dead time, even if it arrives within the dead time of the previous event and is not counted. This adds to the total dead time of the system, whereby a paralyzable system can become totally unresponsive to process events if the count rate of the source is very high. On the other hand, in nonparalyzable systems, the instrument remains insensitive to successive events for a period of time equal to the dead time, and these events are lost. But unlike paralyzable systems, the dead time is not changed or lengthened. When the system recovers after the detection of the first event, only then is the second event processed and detected. The two types of dead time losses are illustrated in Figure 8.9.

The GM counters with quenching gas behave like a nonparalyzable system, whereas most counters including well counters and scintillation cameras are paralyzable systems. In reality, scintillation cameras and rectilinear scanners have both paralyzable and nonparalyzable components of dead time.

Dead-time loss is a serious problem for a counting system at high count rates. Therefore, either count rates must be lowered or corrections must be made to the observed count rates. Methods of determining the dead time correction factors have been detailed in several reference books and readers are referred to them.

Gamma Well Counters

The gamma well counter consists of a NaI(Tl) detector with a hole in the center for a sample to be placed and associated electronics such as a PM tube, preamplifier, amplifier, PHA, and scaler-timer. Placing a radioactive sample in the central hole of the detector increases the geometric efficiency

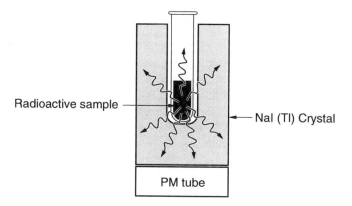

Fig. 8.10. A schematic diagram of a NaI(Tl) well counter. PM, photomultiplier tube.

(almost 99%) and hence the counting efficiency of the counter. The NaI(Tl) detectors have dimensions in the range of 5-cm diameter × 5-cm thick to 23-cm diameter × 23-cm thick. Smaller detectors are used for low-energy γ-rays (less than 200 keV), and larger detectors are used for high-energy γ-rays. Most well counters are shielded with about 8.5-cm thick lead to reduce background from cosmic rays, natural radioactivity such as ^{40}K, or background activity in the work area. A typical well counter detector is shown in Figure 8.10.

It is essential that the dial settings of the discriminators on the PHA are calibrated so that the dial readings correspond directly to the pulse height (i.e., the energy of the γ-ray photon); that is, the dial readings can be read in units of keV. This calibration is called the *high-voltage* or *energy calibration* of the well counter. The energy calibration is carried out by using the 662-keV photons of ^{137}Cs. After placing a ^{137}Cs source in the well counter, the lower and upper discriminator levels are set at 640 divisions and 684 divisions, respectively, thus assigning the center of the photopeak at 662 divisions corresponding to the 662-keV γ-ray. Starting from low values, the high voltage is increased in small increments for a given amplifier gain until the observed count rate reaches a maximum. The discriminator dials are then said to be energy calibrated; for example, each dial unit corresponds to 1 keV. The amplifier gain and the high voltage at the maximum count rate are kept constant for subsequent counting of photons of different energies using the corresponding discriminator settings. For example, the center of the 140-keV photopeak of ^{99m}Tc can be set at 140 divisions of the discriminator setting, with lower and upper values set as desired.

The sample volume affects the counting efficiency of well counters. As the sample volume for a given activity is increased, more radiations are lost through the opening of the well without interacting in the detector, and hence, the counting efficiency drops. Therefore, correction factors should be determined for different sample volumes and applied to the measured activity.

All well counters must be checked regularly for any voltage drift using a long-lived source, such as ^{137}Cs. For relative comparison of count rates between samples, the counter does not need to be calibrated, provided all samples for comparison have the same volume. In radioimmunoassays, ferrokinetics, blood volume, red cell mass measurements, a standard of the same geometry (volume) and relative activity is counted along with all samples, and then a comparison is made between each sample and the standard. However, when the absolute activity of a radioactive sample needs to be determined, then the detection efficiency must be measured for the γ-ray energy of interest using a standard of the radioactive sample of known activity. The photopeak efficiency is determined from the count rate of the standard at appropriate settings divided by the disintegration rate from the known activity of the standard. The efficiency correction can then be applied to the count rates of samples of unknown radioactivity when counted at the same settings as the standard to give the absolute activity.

When multiple γ-rays, either from a single radionuclide or from many radionuclides, are present in a radioactive sample, then the energy spectrum becomes complicated by the overlapping of different photopeaks and also Compton contributions from the high-energy photons to the low-energy photopeaks. Such contributions are termed the *spillover*, or *crosstalk contributions*.

Figure 8.11 illustrates an energy spectrum of the 140-keV peak of ^{99m}Tc and 364-keV peak of ^{131}I, in which the Compton contribution from the 364-keV peak to the 140-keV peak is shown. Corrections must be made for

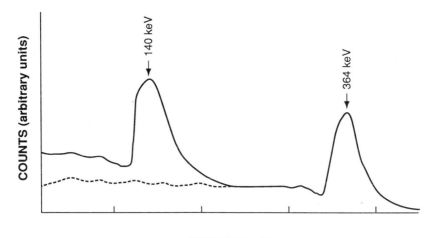

ENERGY (keV)

Fig. 8.11. A combined spectrum of the 140-keV γ-ray of ^{99m}Tc and 364-keV γ-ray of ^{131}I. The dotted line under the 140-keV photopeak is the spillover, or crosstalk, contribution from the 364-keV photon.

this spillover to the 140-keV peak. This is accomplished by counting a sample of pure ^{131}I in both 140-keV and 364-keV discriminator settings and determining the percentage of spillover from the ratio of the counts in the 140-keV photopeak to those in the 364-keV photopeak.

Well counters are available with automatic sample changers having provisions of counting as many as 500 samples. Most counters are programmable with computers and provide printouts with various information on counting. The major advantage of the well counter is its high detection efficiency that is due to increased geometric efficiency which approaches almost 100% depending on the volume of the sample. The detection efficiency of a well counter decreases with increasing photon energy and decreasing detector size. Typically, the overall detection efficiency is 50% to 70% for 140-keV photons of ^{99m}Tc depending on the detector size.

Thyroid Probe

The thyroid probe is a counter commonly employed to measure the uptake of ^{131}I in the thyroid gland after the oral administration of a ^{131}I-NaI capsule. It consists of a NaI(Tl) detector, 5 cm in diameter by 5 cm in thickness, and other associated electronics, as in a well counter. The operation of the probe is similar to that of a well counter.

One of the differences between the well counter and the thyroid probe is that the latter requires a collimator, which limits the field of view on the thyroid. The collimator is a 20- to 25-cm long cylindrical barrel made of lead and covers the detector as well as the PM tube (Fig. 8.12). This prevents the γ-radiations from other organs from reaching the detector.

The efficiency of a probe varies inversely with the distance between the detector and the thyroid. The probe is initially calibrated for photon energies

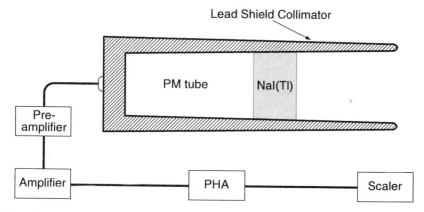

Fig. 8.12. A schematic diagram of a thyroid probe. PHA, pulse-height analyzer; PM, photomultiplier.

in the same manner as the well counter using the 662-keV γ-ray energy of ^{137}Cs, and then discriminator settings are set for the 364-keV γ-ray of ^{131}I. Attenuation of photons in the thyroid tissues reduces the overall detection efficiency of the probe.

Photons scattered in the thyroid gland by Compton scattering may reach and interact in the detector because they originate in the field of view and are not stopped by the collimator thickness. These scattered photons, however, are excluded from the total measured counts by selecting the appropriate lower and upper discriminator settings on the PHA for the 364-keV γ-ray of ^{131}I.

In the thyroid uptake test, a ^{131}I–NaI capsule containing about 10 to 15 μCi (0.37–0.55 MBq) of ^{131}I is measured in a lucite thyroid phantom at a fixed distance using the thyroid probe and the settings for 364-keV photons of ^{131}I. The thickness and composition of the lucite phantom are equivalent to those of the thyroid gland. This count is considered as the standard count. The capsule is then administered to the patient orally, and the thyroid count is obtained at the same distance as the standard count 24 hr after administration. The room background count is taken to subtract from the standard count, and the thigh count is taken as background to subtract from the thyroid count. The thyroid uptake is then calculated as follows:

$$\% \text{ uptake} = \frac{(A - B) \times 100}{(C - D)} \tag{8.7}$$

where A is the thyroid count, B is the thigh count, C is the standard count corrected for 24-hr decay, and D is the room background.

Questions

1. Describe the mechanism of γ-ray interaction in the NaI(Tl) detector. In γ-ray counting, why is NaI(Tl) commonly chosen as the detector?
2. (a) Describe the operation of a photomultiplier (PM) tube.
   (b) What is the typical high voltage applied to the PM tube?
   (c) What are the photocathodes commonly made of?
   (d) How many photoelectrons are emitted from the photocathode for each keV of photon energy?
3. (a) Ideally, a photopeak should appear as a line in a γ-ray spectrum. Indicate different factors that contribute to the broadening of the photopeak.
   (b) A photopeak is due to only photoelectric effect of γ-rays, or due to all γ-rays that deposit full energy in the detector. True or False?
4. (a) Describe the function of a pulse-height analyzer (PHA).
   (b) Do the following factors affect the size (pulse height) of the photopeak pulses?

  i. Gain of the amplifier

  ii. High voltage of PM tubes

  iii. Distance between the source and the detector

  iv. Light photons produced in the detector

5. Describe the origins of:
   (a) Backscatter peak
   (b) Compton valley
   (c) Characteristic K x-ray peak
   (d) Iodine escape peak
   (e) Sum peak in a γ-ray spectrum

6. (a) Describe the principles of a liquid scintillation counter.
   (b) What is a scintillation solution, and how does it work?
   (c) What is the purpose of using a secondary solute to the scintillation solution?
   (d) What are the most common solvents for liquid scintillation counting?
   (e) Can you count ^{3}H (β^{-} energy = 0.018 MeV) and ^{14}C (β^{-} energy = 0.156 MeV) in the same sample using a liquid scintillation counter equipped with three PHAs?

7. (a) Define the energy resolution of a detector.
   (b) For a given detector, the energy resolution of low-energy photons is poorer than that of high-energy photons. True or False?
   (c) For NaI(Tl) detectors, the energy resolution should be less than 10% for the 662-keV photon of ^{137}Cs. True or False?

8. (a) A point source of ^{99m}Tc is placed 10 cm away from a NaI(Tl) detector that has a diameter of 20 cm. Calculate the geometric efficiency.
   (b) What would be the geometric efficiency if the source were placed on the surface of the detector?

9. (a) Explain the dead time of a counter.
   (b) What is the distinction between the paralyzable and nonparalyzable systems?
   (c) What are the typical dead times for Geiger–Müller (GM) counters and NaI(Tl) counters?

10. (a) Describe the energy calibration of a NaI(Tl) well counter.
    (b) Why does the count rate differ from the disintegration rate of a sample of a radionuclide?
    (c) How does the sample volume affect the count rate?

11. What are the spillover or crosstalk contributions in a spectrum of several γ-rays? How would you correct for them?

12. A radioactive sample has two γ-ray photons of 130- and 120-keV energies. If a NaI(Tl) crystal has an energy resolution of 10% at 125 keV, could the two photons be detected as separate photopeaks?

13. Both gas-filled detectors and semiconductor detectors operate by ionization of atoms by radiation. Why do semiconductor detectors give better energy resolution than gas-filled detectors?

14. A patient is given orally a 10-μCi ^{131}I-NaI capsule. Before administration, the count rate of the capsule in a thyroid phantom is 297,000 cpm. The 24-hour count rate of the patient's thyroid is 168,000 cpm. If the room background is 200 cpm and the patient's thigh count rate is 1000 cpm, calculate the thyroid uptake.

15. High-activity sources such as radiopharmaceutical dosages and x-ray exposure outputs are better measured with ionization chambers than GM counters and NaI(Tl) well counters. Why?

16. Which of the following counters can detect individual events of the radiation interacting with the detector?
    (a) Ionization chamber
    (b) GM counter
    (c) NaI(Tl) well counter

17. What type of Compton scattering causes the Compton edge of a γ-ray spectrum?

Suggested Readings

Cradduck TD. Fundamentals of scintillation counting. *Semin Nucl Med.* 1973; 3: 205–223.

Hendee WR. *Medical Radiation Physics.* 2nd ed. Chicago: Year Book Medical Publishers, Inc; 1979.

Hine GJ. Sodium iodide scintillators. In: Hine GJ, eds. *Instrumentation in Nuclear Medicine.* New York: Academic Press; 1967; I: 95–117.

Peng CT, Horrocks DL, Alpen EL, eds. *Liquid Scintillation Counting.* New York: Academic Press; 1980, I, II.

Rollo FD, ed. *Nuclear Medicine Physics, Instrumentation and Agents.* St. Louis: Mosby; 1977.

Sorensen JA, Phelps ME. *Physics in Nuclear Medicine.* 2nd ed. New York: Grune & Stratton; 1987, Chapters 12 and 13.

CHAPTER 9

Imaging Devices

The principles of nuclear medicine studies are based on the assessment of radionuclidic distribution in different parts of a given organ after in vivo administration of a radiopharmaceutical to distinguish between the normal and abnormal tissues. Such assessment of radionuclide distribution is performed by imaging systems that primarily use NaI(Tl) detectors and the associated electronics described in the previous chapter. There are two major imaging systems: rectilinear scanners and scintillation cameras. Rectilinear scanners have been almost phased out in nuclear medicine, whereas scintillation cameras are the equipment of choice for routine use in nuclear medicine imaging. Because of their demise from practical use in nuclear medicine, rectilinear scanners will be only briefly described.

Rectilinear Scanners

A rectilinear scanner usually consists of a NaI(Tl) detector and other conventional electronic circuitry, along with a focused collimator (Fig. 9.1). The detector dimensions are about 7.5–20 cm in diameter and about 2–5 cm in thickness.

Collimators made of lead are used in rectilinear scanners to limit the field of view so that extraneous γ-radiations do not reach the detectors. Although single-hole collimators originally were used, multihole focused collimators became the collimators of choice later for reasons of higher efficiency and better spatial resolution. These collimators are constructed with multiple tapered holes converging to a common point outside the collimator, called the *focal point*, or *focus* (Fig. 9.2). Focal length is the distance between the focus and the collimator face. The septal thickness of lead between holes is sufficient to stop photons of certain energies, thus providing improved spatial resolution. Focused collimators are classified as low-energy–high-resolution, low-energy–medium-resolution, medium-energy, and high-energy collimators, and they can be used for radionuclides with different γ-ray energies. The characteristics of the collimators are discussed in Chapter 10.

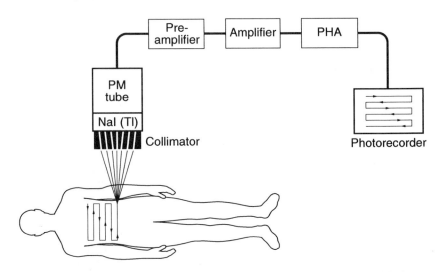

Fig. 9.1. Principles of a rectilinear scanner. The rectilinear scanner (NaI(Tl) detector equipped with a multihole focused collimator) scans the body of a patient from one end to the other at an increment of a small distance. A photorecorder records a dot each time a pulse is generated from γ-ray interaction at a location in the patient's body. PHA, pulse-height analyzer; PM, photomultiplier.

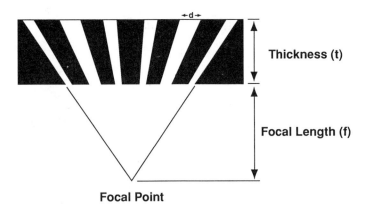

Fig. 9.2. A multihole converging collimator with a focal point. The distance between the focal point and the collimator face is the focal length (f). The collimator has thickness, t, and hole diameter, d.

The principle of rectilinear scanning involves the moving back and forth of the collimated detector across the area of interest on the patient, starting from one end of the area to the other end and advancing only a small but constant distance each time the detector comes to the end of its traverse and changes its direction. The γ-rays from the field of view interact with the NaI(Tl) detector, and pulses produced by the photomultiplier (PM) tube are amplified by an amplifier and analyzed by a pulse-height analyzer (PHA), using appropriate discriminator settings for γ-ray energy in question. The "accepted" pulses are then fed into a scaler-timer to represent counts and to an output display device such as a paper tapper, a photorecorder, or a persistence oscilloscope. Displayed data are synchronized with the motion of the detector so that the X and Y coordinates of the distribution of the activity in the area of interest are accurately defined. Persistence oscilloscopes are used to see the live image on the cathode ray tube (CRT) screen.

Recording is made by a paper tapper by making a dot on the paper each time a pulse is received. This technique was not well received and was later replaced by the photorecorder technique. In the latter technique, a small light source is generated by the pulse and moves over an x-ray film inside a light-tight box in synchronization with the motion of the detector. The light beam is further collimated by a narrow slit and projects a small dot on the film. The amount of darkening and the number of dots per unit area, that is, the density of the image on the film, depend on the count rates per unit area of the object as detected by the detector. Higher count rates produce greater film densities and vice versa.

The quality of an image formed by rectilinear scanning is, in part, defined by the counts information density (CID) or simply information density (ID). Information density is given by

$$\text{ID (counts/cm}^2) = \frac{C}{S \times LS} \tag{9.1}$$

where C is the count rate (counts/min); S is the scan speed (cm/min); and LS is the line spacing (cm) between two passes. The scan quality is improved by increasing the count rate, that is, the administered activity, and by reducing the speed of the detector. The optimal information density for most nuclear medicine studies is about 1000 counts/cm^2.

After positioning the patient with the area of interest under the detector, the maximum count rate in the area is determined by physically moving the detector head by hand over the area. Based on this count rate and the optimum information density desired, the scan speed is set at the appropriate value. The line spacing may be several millimeters at scan speeds of several hundred centimeters per minute.

Rectilinear scanners are also available with dual probes that are connected axially to each other and operate synchronously. The patient is placed between the two probes, which then produce two views simultaneously, such as posterior and anterior or right and left lateral views.

Scintillation or Gamma Cameras

The scintillation camera, or gamma camera, is a stationary imaging device that is most commonly used in nuclear medicine. It is also called the Anger camera in honor of Hal O. Anger, who invented it in the late 1950s. Unlike the rectilinear scanner, cameras detect radiation from the entire field of view simultaneously and therefore are capable of recording dynamic as well as static images of the area of interest in the patient. Thus, imaging time with scintillation cameras is considerably less than that with rectilinear scanners. Various designs of scintillation cameras have been proposed and made available, but the Anger camera with a single crystal is by far the most widely used. Although many sophisticated improvements have been made on the scintillation cameras over the years, the basic principles of the operation are essentially the same.

Principles of Operation

The gamma camera usually consists of several components: a detector, a collimator, PM tubes, a preamplifier, an amplifier, a PHA, an X-, Y-positioning circuit, and a display or recording device. A schematic diagram of a gamma camera is illustrated in Figure 9.3, and a gamma camera is shown in Figure 9.4. The detector, PM tubes, and amplifiers are housed in a unit called the *detector head*, which is mounted on a stand. The head can be moved up or down and rotated with electrical switches to position it in the field of view on the patient. The X-, Y-positioning circuits, PHA, and some recording devices are mounted on a console. In the past, the cameras were operated by switches and dials on the console. Nowadays, the operation of the camera is performed by a computer built in it. The computer is run by appropriate software packages in conjunction with a keyboard and a video monitor (see Fig. 9.4). High voltage, intensity, window, and photopeaks are all set by the operator's choice of parameters in the computer. Often, acquisition of the data and, at times, processing of the data are carried out by

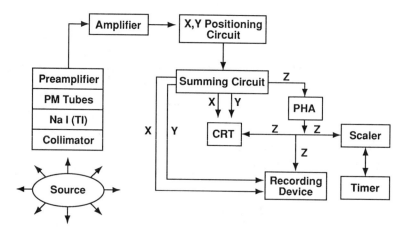

Fig. 9.3. A schematic electronics diagram of a gamma camera. CRT, cathode ray tube; PHA, pulse-height analyzer; PM, photomultiplier.

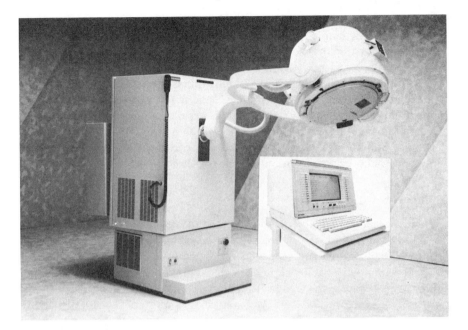

Fig. 9.4. A typical gamma camera. (Courtesy of Siemens Medical Systems, Inc.)

the computer, depending on the manufacturer. Whereas stationary cameras are permanently installed at desired locations, mobile gamma cameras are mounted on wheels for use in situations requiring movement of the camera from room to room, such as to the patient's bedside.

The operational principles of a gamma camera are identical to those of solid scintillation counters described in Chapter 8. Basically, γ-rays from a source interact with the NaI(Tl) detector, and light photons are emitted. The latter strike the photocathode of PM tubes, and a pulse is generated, which is then amplified by an amplifier and sorted out by a PHA. Finally, the pulse is positioned by an X-, Y-positioning circuit on the recording device corresponding to the location of γ-ray interaction in the detector.

The functions of PM tubes, preamplifier, amplifier, PHA, and recording devices are the same as described in Chapter 8, and therefore only essential features pertaining to the use of gamma cameras will be highlighted.

Detector

Detectors used in gamma cameras are typically circular NaI(Tl) detectors, which have dimensions of 25–50 cm in diameter and 0.64–1.27 cm in thickness. The most common thickness is 0.95 cm. The 0.64-cm thick detectors are used in mobile gamma cameras and are useful for ^{201}Tl, ^{99m}Tc, and ^{123}I radionuclides. Larger crystals ($>$ 40 cm in diameter) are used in large field of view (LFOV) cameras. Rectangular NaI(Tl) detectors (38.7 × 61 cm or 45 × 66 cm) are also available in gamma cameras.

Increasing the thickness of a detector increases the probability of complete absorption of γ-rays and hence the sensitivity (defined in Chapter 10) of the detector. However, the probability of multiple Compton scattering also increases in thicker detectors, and therefore the X, Y coordinates of the point of γ-ray interaction can be obscured. This results in poor resolution of the image of the area of interest. For this reason, thin NaI(Tl) detectors are used in gamma cameras, but this decreases the sensitivity of the camera, because many γ-rays may escape from the detector without interaction.

Collimator

In gamma cameras, a collimator is attached to the face of the NaI(Tl) detector to limit the field of view so that γ-radiations from outside the field of view are prevented from reaching the detector. Collimators are normally made of material with high atomic number, such as tungsten, lead, and platinum, among which lead is the material of economic choice in nuclear medicine. They are designed in different sizes and shapes and contain one or many holes to view the area of interest. The collimator with one hole is called a *pinhole collimator*. Parallel-hole collimators have between 4000 and 46,000 holes depending on the collimator design.

Classification of collimators depends primarily on the type of focusing and also the septal thickness of the holes. Depending on the type of focusing, collimators are classified as parallel-hole, pinhole, converging, and diverging types; these are illustrated in Figure 9.5. Pinhole collimators are made in conical shape with a single hole and are used in imaging small organs such as the thyroid glands to provide magnified images. Converging collimators are made with holes converging to an outside point and are employed to provide magnified images when the organ of interest is smaller than the size of the detector. A typical converging collimator is a fan beam collimator commonly used in brain imaging. Diverging collimators are constructed with holes that

COLLIMATOR DESIGNS

Parallel hole collimator   Diverging collimator

Pinhole collimator   Converging collimator

Fig. 9.5. Different designs of collimators.

are divergent from the detector face and are used in imaging organs such as lungs that are larger than the size of the detector. The organ images are minified with these collimators. Parallel-hole collimators are made with holes that are parallel to each other and perpendicular to the detector face. These collimators are most commonly used in nuclear medicine procedures and furnish a one-to-one projected image. Because pinhole and converging collimators magnify and the diverging collimators minify the image of the object, some distortion occurs in images obtained with these collimators. Because LFOV cameras are readily available now, diverging collimators are seldom used.

Parallel-hole collimators are classified as high-resolution, all-purpose, and high-sensitivity type, or low-energy, medium-energy or high-energy type, depending on the resolution and sensitivity they provide in imaging. These characteristics are discussed in detail in Chapter 10.

Photomultiplier Tube

As in scintillation counters, PM tubes are essential in gamma cameras for converting the light photons in the NaI(Tl) detector to a pulse. Instead of one PM tube, an array of PM tubes (19 to 91) are mounted in a hexagonal fashion to the back of the detector with optical grease, or in some instances, using lucite light pipes between the detector and the PM tubes. In modern gamma cameras, square or hexagonal PM tubes are used for better packing. The output of each PM tube is used to define the X, Y coordinate of the point of interaction of the γ-ray in the detector by the use of an X-, Y-positioning circuit (see later) and also is summed up by a summing circuit to form a pulse known as the Z *pulse*. The Z pulse is then subjected to pulse-height analysis and is accepted if it falls within the range of selected energies.

X-, *Y*-Positioning Circuit

Each pulse arising out of the γ-ray interaction in the NaI(Tl) detector is projected at an X, Y location on the image corresponding to the X, Y location of the point of interaction of the γ-ray. This is accomplished by an X-, Y-positioning circuit in conjunction with the PM tubes and a summing circuit. Figure 9.6 illustrates the principles of X, Y positioning of pulses arising from γ-ray interactions in the detector employing seven PM tubes. All PM tubes are connected through capacitors to four output leads representing four directional signals, X^+, X^-, Y^+, and Y^-. The capacitance values are assigned in direct proportion to the location of the PM tube relative to the four signals. Suppose a γ-ray interacts at a location (*) near tube 7. The largest amount of light is received by tube 7, and other tubes receive light in proportion to their distances from the point of interaction. The output signals of PM tubes are weighted by the appropriate capacitance values and then summed to form each of the X^+, X^-, Y^+, and Y^- signals individually. In this

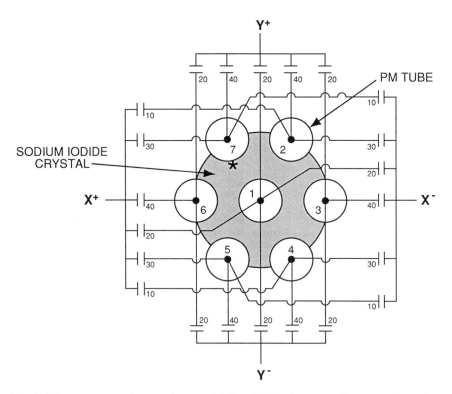

Fig. 9.6. Arrangement of seven photomultiplier (PM) tubes to produce X and Y pulses for the X, Y location of the γ-ray interaction in the detector. X^+, X^-, Y^+, and Y^- pulses are obtained by summing the output of all PM tubes weighted by capacitors for the location of each PM tube in relation to the site of γ-ray interaction. (Adapted from Anger HO. Scintillation camera. Rev Sci Instr. 1958;29:27.)

case, X^+ will be greater than X^-, and Y^+ will be greater than Y^-, because the interaction occurred in the upper left quadrant. The X-, Y-designating pulses, X and Y, and the Z pulse are then obtained as follows:

$$Z = X^+ + X^- + Y^+ + Y^- \tag{9.2}$$

$$X = \frac{k}{Z}(X^+ - X^-) \tag{9.3}$$

$$Y = \frac{k}{Z}(Y^+ - Y^-) \tag{9.4}$$

where k is a constant and k/Z is the amplifier gain. The X and Y pulses are then projected on a CRT to depict the X, Y coordinates of the point of γ-ray interaction, which in turn corresponds to the coordinates of the location in the field of view from which the γ-ray originated. Similarly, these pulses can be stored in the computer in a square matrix so that the data can be processed later to reproduce an image. Or, they can be projected on x-ray film.

The larger the number of PM tubes, the better the accuracy of the X, Y locations of all pulses on the image; that is, the better the spatial resolution on the image (see Chapter 10).

Pulse-Height Analyzer

After the Z pulses are formed by the summing circuit, the PHA analyzes their amplitude and selects only those of desired energy by the use of appropriate peak and window settings. In many gamma cameras, the energy selection is made automatically by push-button type isotope selectors designated for different radionuclides such as ^{99m}Tc, ^{131}I, and so on. In modern cameras, isotope peak and window settings are selected by the mouse-driven menu on a computer interfaced with the camera. In some gamma cameras, two or three PHAs are used to select simultaneously two or three γ-rays of different energies. These types of cameras are useful in imaging with ^{111}In and ^{67}Ga that possess two or three predominant γ-rays. The window settings are provided in percentages of the peak energy by a control knob. For most studies, a 20% window centered symmetrically on the photopeak is employed.

It should be pointed out that X and Y pulses are applied to a CRT or recorded on films only if the Z pulse is within the energy range selected by the PHA. If the Z pulse is outside this range, then X and Y pulses are discarded.

Display and Recording Systems

The pulses after pulse height analysis and X, Y positioning are projected on a persistence oscilloscope to see the instantaneous image of the distribution of activity in a source. Data are either collected for a preset time or for preset total counts (e.g., 500,000 counts) using a scaler-timer. Images are normally obtained on photographic films (35- and 70-mm transparency film, x-ray film, or Polaroid film) by accumulating data over a period of time. The data can be stored on a magnetic tape or disk or in a computer for processing later.

Polaroid films have the advantage of rapid development in 15–20 sec so that repeat images can be taken immediately, if necessary. But they are expensive, have poor gray scale and contrast, and are not suitable for sequential dynamic studies. Transparency roll films are used with cameras having an automatic advancing mechanism for recording images in rapid sequences. X-ray films are employed in multiformat systems in which one to 64 images are recorded on a single film (usually 20 × 25 or 28 × 36 cm), thus providing many views in different projections for comparison. Optimal contrast of the image is obtained by properly adjusting the intensity control knob on the console of the image recording device, which is usually a separate unit from the camera.

Digital Computer

Over the years, the use of digital computers in nuclear medicine has considerably increased, and more and more nuclear medicine studies are being analyzed by computers. Data are stored and images are formed and displayed on the video monitor—all by digital computers.

The signals from a gamma camera are in analog form and must be digitized before processing by the computer. Most digital computers operate with binary numbers, and the basic information unit of the computer memory is a bit (*bi*nary digi*t*). Binary digits are expressed in powers of 2, as opposed to decimal numbers, which are expressed in powers of 10. The X- and Y-analog signals are digitized (converted into bits) by a device in the camera–computer interface, known as the analog-to-digital converter (ADC). The ADC can be 8- to 16-bit type and accordingly converts the range of X and Y pulses into bits that are in proportion to the X, Y coordinates of the site of interaction of the photons in the detector. The bit capacity of the ADCs affects the spatial resolution of the image and is mostly dictated by the hardware design.

The computer memory approximates the area of the detector as a square matrix of a definite size, most commonly 64×64 and 128×128 in nuclear medicine with 4096 (4K) and 16,384 (16K) picture elements, called pixels, respectively. The choice of a matrix of a given size is made by the operator, depending on the type of software used. Each pixel corresponds to a specific X, Y location in the detector. How many events or counts can be registered in a given pixel depends on the depth of the pixel, which is given by a *byte* or a *word*. A byte is 8 bits long and can store counts up to $2^8 - 1 = 255$ (the number 1 is subtracted because 0 is considered a count in binary systems). A word is 16 bits long and can accommodate counts up to $2^{16} - 1 = 65,535$.

The binary digits resulting from the ADC conversion of the X and Y pulses are stored in the appropriate pixel of the chosen matrix. The faster the ADC, the higher the count rate that can be counted. The slower ADCs increase the dead time of the system. As the counting progresses, more counts are accumulated in appropriate locations, and finally a digitized image is stored corresponding to the radionuclidic distribution in the patient.

Data are collected in either the frame mode or the list mode. The frame mode is the most common method in nuclear medicine, in which the digitized data obtained from the X and Y pulses are stored in the corresponding X, Y positions of the matrix of choice in the computer. In this mode, one must specify the matrix size desired; the number of images (frames) per study; time of collection of data per frame or total counts to be collected per frame. This mode of data acquisition furnishes instant images for storage and display.

In the list mode acquisition, digitized data from X and Y pulses are stored and coded as they are received in sequence in time. After the study is completed, the data are then reformatted into frames as in the frame mode according to the choice of the matrix and framing rate. The disadvantages of this method are that processing takes a long time and that images are not available soon after the completion of the study.

Multicrystal Cameras

In the early 1960s, Bender and Blau introduced a multicrystal camera, called the Autofluoroscope, which consists of 294 NaI(Tl) detectors, each 0.8 cm in diameter and 3.8 cm in thickness, arranged in a 14 × 21 array. The unit is furnished with 35 PM tubes, instead of 294, each row and each column being coupled to a separate single PM tube via a light pipe. The principles of operation are the same as those of the Anger camera except that the X, Y positioning is simplified. This system was later marketed as System 70.

The multicrystal camera is mostly used for fast, dynamic imaging, although static imaging also can be performed. High count-rate capability and greater high-energy photon detection efficiency are the major advantages of these cameras. The energy resolution of these cameras, however, is poorer than that of the Anger cameras, which results in poorer contrast on the image.

The latest version of the multicrystal camera is Scinticor, which uses a single sheet of NaI(Tl) detector. Grooves are made on the detector such that an array of columns and rows is created on which PM tubes are placed, thereby increasing both counting efficiency and energy resolution.

Multicrystal cameras are expensive and not commonly used in nuclear medicine.

Questions

1. (a) Describe the operation of a rectilinear scanner.
   (b) An optimum information density of 1000 counts/cm^2 is needed in most nuclear medicine images. If the maximum count rate over the area of interest is 39,000 cpm, and the line spacing for the rectilinear scanner is 0.3 cm, what is the scan speed needed?
   (c) What type of collimator is used in rectilinear scanning?
2. (a) Describe the operational principles of a gamma camera.
   (b) The main purpose of a collimator is to limit the field of view of an imaging device for imaging. True or False?
   (c) The purpose of a photomultiplier tube is to convert light photons to an electron pulse. True or False?
   (d) The scattered photons are excluded by the proper choice of a collimator. True or False?
   (e) Scattered photons are excluded by the proper choice of discriminator settings (windows). True or False?

3. (a) What are the different categories of collimators?
   (b) Which collimator is most used in nuclear medicine?
   (c) Which types of collimator give image distortion and why?
4. Describe the function of the X, Y circuit in the gamma camera system.
5. A pulse-height analyzer:
   (a) Reduces the background. True or False?
   (b) Rejects γ-rays that undergo Compton scattering in the patient and the detector. True or False?
   (c) Rejects γ-rays undergoing photoelectric effect in patients. True or False?
   (d) Increases the signal-to-noise ratio. True or False?
6. (a) The counting efficiency of a gamma camera increases with the thickness of the detector. True or False?
   (b) What are the most common thicknesses of the NaI(Tl) detector used?
   (c) A gamma camera detector with a 20-cm field of view is used to image the lungs, which fill 75% of the image. The camera is set to accumulate 450,000 counts. Calculate the information density.
7. Discuss the differences between the rectilinear scanner and the gamma camera.
8. Describe how a digital computer works when interfaced with a gamma camera.
9. In pulse-height analysis, a 20% window means 10% on either side of the photopeak. True or False?
10. What are the differences between the frame mode and list mode collections of data in a computer?

References and Suggested Readings

Bender MA, Blau M. The autofluoroscope. *Nucleonics.* 1963;21:52.

Erickson J. Imaging systems. In: Harbert J, da Rocha AFG, eds. *Textbook of Nuclear Medicine, Volume I: Basic Science.* Philadelphia: Lea & Febiger; 1984.

Rollo FD, ed. *Nuclear Physics, Instrumentation, and Agents.* St Louis: CV Mosby; 1977.

Sorensen JA, Phelps ME. *Physics in Nuclear Medicine.* 2nd ed. New York: Grune & Stratton; 1987.

CHAPTER 10

Performance Parameters of Imaging Devices

The quality and detailed display of an image are affected by several parameters associated with the imaging systems. These parameters include spatial resolution, sensitivity, uniformity, and contrast, and their various features are described here in detail. A brief description of the quality control tests for gamma cameras is also included.

Spatial Resolution

Gamma Camera

The spatial resolution of an imaging device is a measure of the ability of the device to faithfully reproduce the image of an object, thus clearly depicting the variations in the distribution of radioactivity in the object (Erickson, 1984). The *spatial resolution* of an imaging system is empirically defined as the minimum distance between two points in an image that can be detected by the system. The overall spatial resolution (R_o) of an imaging device comprises three components, namely, intrinsic resolution (R_i) of the detection system, collimator resolution (R_g), and scatter resolution (R_s), and is given by

$$R_o = \sqrt{R_i^2 + R_g^2 + R_s^2} \qquad (10.1)$$

The smaller numerical values of R$_o$ indicate better resolution and vice versa.

Intrinsic Resolution

Intrinsic resolution, R_i, is the component of spatial resolution contributed by the detector and associated electronics, and is a measure of how well an imaging device can localize an event. Intrinsic resolution arises primarily from the statistical fluctuations in pulse formation that have been discussed in the section entitled Gamma Ray Spectrometry in Chapter 8. The statistical fluctuations include variations in the production of light photons after γ-ray interaction in the detector and variations in the number of electrons emitted from the photocathode and dynodes in the photomultiplier (PM) tubes. In gamma cameras, the X, Y positioning of the pulses is improved by increasing the number of PM tubes, thus improving the intrinsic resolution. Also, PM tubes with greater sensitivity are used and improved optical coupling is employed to attach them to the detector for greater light collection, thus leading to better intrinsic resolution.

Intrinsic resolution improves with higher γ-ray energy and deteriorates with lower energy photons, because greater statistical fluctuations occur in the production of light photons by lower energy photons and vice versa. Intrinsic resolution improves with narrow PHA window settings, because scattered radiations are avoided.

Multiple Compton scattering of a γ-ray photon in the detector resulting in the absorption of the total photon energy causes uncertainty in the X, Y location of the original γ-ray interaction and makes the intrinsic resolution worse. This effect is worse with thicker detectors because of the increased chances of multiple scattering. For this reason, only thinner detectors (0.63–1.25 cm) are used in gamma cameras.

Collimator Resolution

Collimator resolution, also termed the *geometric resolution* (R_g), constitutes the major part of the overall spatial resolution and primarily arises from the collimator design. As already mentioned in Chapter 9, there are four types of collimators: parallel-hole, pinhole, converging, and diverging. Of these, parallel-hole collimators are most commonly used in nuclear medicine.

A typical parallel-hole collimator is shown in Figure 10.1. The spatial resolution for this collimator is given by the geometric radius of acceptance, R_g:

$$R_g = \frac{d(t_e + b + c)}{t_e} \tag{10.2}$$

where d is the hole diameter of the collimator, b is the distance between the collimator face and the source of radiation, c is the distance between the back face of the collimator and the midplane of the detector, and t_e is the effective length of the collimator holes. The t_e is empirically given by $t_e = t - 2\mu^{-1}$, where μ is the linear attenuation coefficient of the collimator material (e.g., lead), and t is the length or thickness of the collimator hole. This corrects for the penetration of the two corners of the holes by the photons.

As seen from Eq. (10.2), the collimator resolution is improved by increasing the length, t, of the collimator holes or by decreasing the diameter, d, of the holes. Thus, long narrow holes provide best spatial resolution. Also, the collimator resolution deteriorates with increasing source-to-collimator distance, b, and is best at the collimator face. Therefore, in nuclear medicine studies, patients should be placed as close to the collimator as possible to provide the best resolution.

The thickness of lead between holes is called the septum (a). *Septal penetration* of γ-rays plays an important role in the collimator resolution and de-

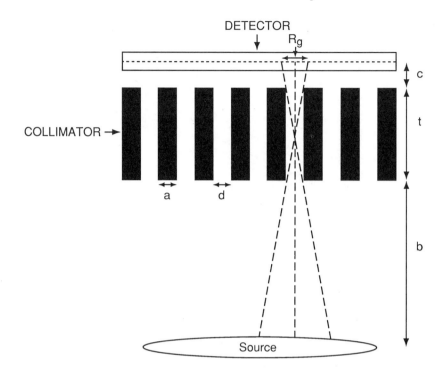

Fig. 10.1. A parallel-hole collimator with thickness t, hole diameter d, septal thickness a, and source-to-collimator distance b. The collimator is attached to a detector whose midplane is at a distance c from the back surface of the collimator. R_g is the collimator resolution.

pends on the γ-ray energy. High-energy photons from outside the field of view can cross the septum and yet interact in the detector, thus obscuring the image. Because of this, γ-rays of only ~50–300 keV are suitable for present-day collimators, the most preferable photon energy being 150 keV. At energies below ~50 keV, photons are absorbed in the body tissue, whereas at energies above ~300 keV, septal penetration of the photons can occur. Current collimators are made with appropriate septal thickness for specific photon energies to limit septal penetration. Parallel-hole collimators are classified as low-energy collimators with a few tenths of a millimeter septal thickness (for up to 150-keV γ-rays) and medium-energy collimators with a few millimeter thickness (up to 450-keV photons) (Sorensen and Phelps, 1987). It is understandable that for a given diameter collimator, the number of holes are greater in low-energy collimators than in high-energy collimators. Normally, high-energy collimators have poorer efficiency and resolution than low-energy collimators.

In another classification, collimators are termed high-sensitivity and high-resolution collimators. Often, these collimators are made with an identical number of holes with identical diameters but with different thicknesses. Thus,

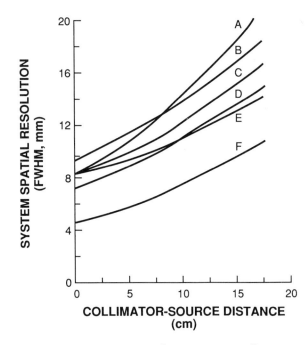

Fig. 10.2. Effect of source-to-collimator distance on overall system resolution for various types of collimators. (A) High sensitivity parallel hole. (B) Diverging. (C) All purpose parallel hole. (D) Converging. (E) High resolution parallel hole. (F) Pinhole. (From Rollo FD, Harris CC. Factors affecting image formation. In: Rollo FD, ed. *Nuclear Medicine Physics, Instrumentation and Agents*. St. Louis: Mosby; 1977:407. Modified from Moyer RA. J Nucl Med 1974, 15:59.)

the collimator with longer holes is called the high-resolution collimator and that with shorter holes is called the high-sensitivity collimator. The spatial resolution for the high-sensitivity collimator deteriorates sharply with the source-to-collimator distance. All-purpose, or general-purpose, collimators are designed with intermediate values of resolution and sensitivity.

The collimator resolution for pinhole, diverging, and converging collimators is expressed by similar but somewhat complex equations, and their details are available from reference books on nuclear physics and instrumentation. For pinhole and converging collimators, best resolution is obtained when the object is at the focal plane. The overall system resolutions of different collimators are illustrated in Figure 10.2.

A recent collimator called the *fan beam collimator* is basically a converging collimator and commonly is used for brain imaging. It gives better spatial resolution but poorer sensitivity than parallel-hole collimators.

Scatter Resolution

Radiations are scattered in patients. It is possible that radiations that are not within the field of view may be scattered in the body without much loss of energy and fall within the field of view. Scattered radiations can interact with the detector and result in pulses of acceptable amplitude set by pulse-height analyzer (PHA). This degrades the overall spatial resolution. This component is called the *scatter resolution* (R_s) and depends on the composition of the source of radiation and PHA discriminator settings. The effect of scatter resolution is essentially the same for all collimators (Rollo and Harris, 1977).

Rectilinear Scanner

Because the rectilinear scanners are operated without X-, Y-positioning circuits, the intrinsic resolution, R_i, is not considered an essential component of overall resolution for these imaging devices. Thus, from Eq. (10.1), the overall spatial resolution, R_o, for rectilinear scanners is

$$R_o = \sqrt{R_g{}^2 + R_s{}^2} \tag{10.3}$$

where R_g is the collimator resolution and R_s is the scatter resolution. For a focused collimator (see Fig. 9.2), R_g is given by

$$R_g = \frac{d \cdot f}{t} \tag{10.4}$$

where d is the diameter of the hole on the detector side, f is the focal length of the collimator, and t is the thickness of the collimator.

In imaging with rectilinear scanners, best spatial resolution is achieved when the source of radiation (e.g., the patient) is at the focal plane. The spatial resolution is degraded in images obtained at planes above or below the focal plane.

Blur

A term related to spatial resolution is *blur*. Blur results from the broadening of the image of each point source in a distribution of radioactivity (e.g., the patient). The performance characteristics of imaging devices result in blurring. It is understandable that blurring is an inverse function of spatial resolution. That is, spatial resolution deteriorates with increasing blur and vice versa. Blur affects the contrast of the image and depends on the type of imaging device used and the intensity of the radiation source.

Evaluation of Spatial Resolution

Bar Phantom

Qualitative evaluation of the spatial resolution of an imaging device can be made by visual inspection of the images obtained using bar phantoms. Bar phantoms consist of four sets of parallel lead bar strips arranged perpendicular to each other in four quadrants in a lucite holder (Fig. 10.3A). The widths

Fig. 10.3. (A) Bar phantom. (B) Hine–Duley phantom. (Courtesy of Nuclear Associates, Carle Place, NY.)

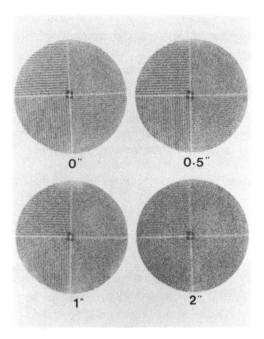

Fig. 10.4. Bar phantom images at different distances from a parallel-hole collimator.

and spacings of the strips are the same within each quadrant but differ in different quadrants. The Hine–Duley phantom consists of five groups of lead strips of different thicknesses and spacings arranged in parallel fashion in a lucite holder (Fig. 10.3B). In all bar phantoms, the thickness of lead should be sufficient to stop photons of a given energy for which spatial resolution is being estimated.

The bar phantom is placed over the detector of a gamma camera. A flood source of equivalent dimension containing sufficient activity (usually, 5 to

10 mCi ^{57}Co, i.e., 185 to 370 MBq ^{57}Co) is placed on the top of it and an image is taken. Cobalt-57 is used because it has a longer half-life of 270 days, compared to 6 hr of ^{99m}Tc, and also has 122- and 136-keV γ-rays equivalent to 140-keV γ-rays of ^{99m}Tc. These flood sources are commercially available in circular or rectangular forms. For evaluation of spatial resolution for different photon energies, flood sources of radiations of respective energies should be used.

The image of the bar phantom obtained by the preceding method is visually inspected, and spatial resolution is estimated from the smallest strips or spacings distinguishable on the image (Fig. 10.4). Obviously, this technique is qualitative and does not give a quantitative measure of the spatial resolution.

Line-Spread Function

An improved method of estimating spatial resolution of an imaging system is based on the use of a line-spread function (LSF). A long plastic tubing filled with a radioactive solution is moved across the field of view of the detector. The counts (response) obtained at incremental distances are plotted against the distance from the center axis of the collimator to give a bell-shaped LSF (Fig. 10.5). A gamma camera interfaced with a computer is often used to collect and store counts from the line source in a single view, and then the computer generates the LSF. The full width at half maximum (FWHM) of the LSF curve gives the spatial resolution of the imaging device.

Spatial resolution by the LSF method varies with the design of the collimator and is, therefore, different for parallel-hole, converging, and diverging collimators. Also, the FWHM values of the LSF may not represent the true spatial resolution, because the scatter and septal penetration components fall in the tail part of the LSF (i.e., below 50%) and therefore are not accounted for.

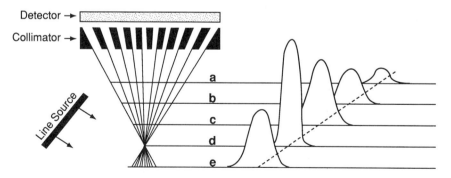

Fig. 10.5. Line-spread functions at different distances a, b, c, d, and e from a focused collimator. The distance d is at the focal point and hence the sharper line-spread function.

Modulation Transfer Function

A more complete and quantitative assessment of spatial resolution of imaging devices is made by using modulation transfer function (MTF) (Rollo, 1977). The concept of MTF is illustrated in Figure 10.6. Suppose a source of activity (e.g., a patient) has a sinusoidal distribution with peaks (maximum activity, A_{max}) and valleys (minimum activity, A_{min}), as illustrated in Figure 10.6. Such a distribution gives a spatial frequency (v) in cycles per centimeter or cycles per millimeter. The contrast, or *modulation* (M_s) in the source activity is given by

$$M_s = \frac{A_{max} - A_{min}}{A_{max} + A_{min}} \tag{10.5}$$

If a perfect imaging device were to image the source faithfully, it would depict the same distribution of activity in the image with A_{max} and A_{min} as in the source activity. Because the imaging devices are not absolutely perfect, it will portray the distribution of activity in the image with C_{max} for the peak and C_{min} for the valley, which are smaller in magnitude than A_{max} and A_{min}. The modulation in the image (M_i) is then expressed by

$$M_i = \frac{C_{max} - C_{min}}{C_{max} + C_{min}} \tag{10.6}$$

The MTF at a spatial frequency v is then calculated as the ratio of M_i to M_s:

$$MTF(v) = \frac{M_i}{M_s} \tag{10.7}$$

When $M_s = M_i$, MTF = 1. This is true if the sinusoidal cycles are well separated and if the imaging device reproduces the image of each cycle faithfully. Thus, the system with MTF = 1 gives the best overall spatial resolution. When the distribution of activity is such that spatial frequency increases, the peaks and valleys come closer. When the peaks and valleys are too close, the imaging device cannot delineate them, and the MTF tends to 0, yielding the

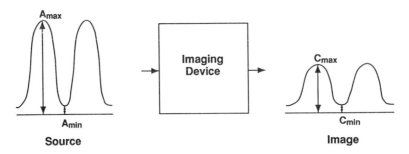

Fig. 10.6. An illustration of the principles of modulation transfer function (MTF) of an imaging system. See text for details.

poorest spatial resolution of the system. The values of the MTF between 0 and 1 give intermediate spatial resolutions.

It has been demonstrated that the MTF is a normalized Fourier transform of the LSF discussed previously. In practice, the source activity distribution is assumed to be composed of line sources separated by infinitesimal distances, and the MTFs are then calculated from the LSFs of all line sources. The mathematical expression of these functions is quite complex and can be found in reference physics books.

Plots of the MTFs against spatial frequencies are useful in determining the overall spatial resolution of imaging devices and are presented in Figure 10.7 for three imaging systems. It is seen that, at lower frequencies (i.e., larger separation of sinusoidal cycles), the MTFs are almost unity for all three systems; that is, they reproduce equally good images of the radiation source. At higher frequencies (i.e., sinusoidal cycles are too close), system A in Figure 10.7 gives the best resolution, followed in order by system B and system C.

It is appropriate to mention that whereas the FWHM of the LSF does not account for the scatter and septal penetration of γ-radiations, the MTF takes these two factors into consideration and provides a complete description of the spatial resolution of a system. Furthermore, individual components of an imaging system may have separate MTFs, and they are combined to give the overall MTF, as follows:

$$MTF = MTF_1 \times MTF_2 \times MTF_3 \ldots \tag{10.8}$$

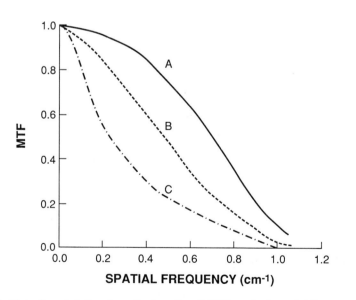

Fig. 10.7. Plot of modulation transfer function (MTF) against spatial frequency. System A gives better spatial resolution than systems B and C, and system B provides better resolution than system C.

Problem 10.1
The MTFs at a certain spatial frequency of the detector, PM tubes, and PHA of a gamma camera are 0.8, 0.6, and 0.7, respectively. What is the overall MTF of the camera?

Answer
$$\text{MTF} = 0.8 \times 0.6 \times 0.7$$
$$= 0.34$$

Sensitivity

Sensitivity of an imaging device is defined as the number of counts per unit time detected by the device for each unit of activity present in a source. It is normally expressed in counts per second per microcurie (cps/μCi). Sensitivity depends on the geometric efficiency of the collimator, the detection efficiency of the detector, PHA discriminator settings, and the dead time. Efficiency of a detector (Chapter 8), PHA discriminator settings (Chapters 8 and 9), and the dead time (Chapter 8) have been discussed previously. Briefly, the counting efficiency of a detector decreases with increasing photon energy and with increasing source-to-detector distance but increases with the thickness of the detector. A narrow window setting on the PHA reduces the counts measured and therefore compromises the counting efficiency. Counts are lost when counting a high-activity sample using a device with a long dead time, and hence the counting efficiency is reduced. Sensitivity of a counting system is primarily affected by the collimator efficiency, which is described next.

Collimator Efficiency

Collimator efficiency, or geometric efficiency (E_g), is defined as the number of γ-ray photons passing through the collimator holes per unit activity present in a source. For parallel-hole collimators (see Fig. 10.1), it is given by

$$E_g = K \cdot \frac{d^4}{t_e^{2}(d + a)^2} \qquad (10.9)$$

where d is the hole diameter, t_e is the effective length of the collimator hole defined before, and a is the septal thickness. The constant K is a function of the shape and arrangement of holes in the collimator.

The collimator efficiency for parallel-hole collimators increases with increasing diameter of the collimator holes and decreases with increasing collimator thickness (t) and septal thickness (a), which is quite opposite to spatial resolution [see Eq. (10.2)]. Thus, *as the spatial resolution of a system increases, its sensitivity decreases, and vice versa for a given collimator*. Note

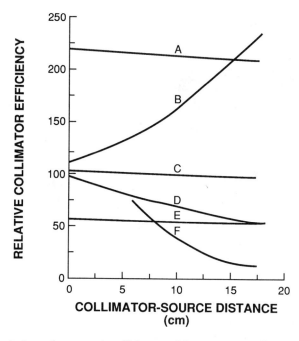

Fig. 10.8. Variation of geometric efficiency with source-to-collimator distance for various collimators. (A) High sensitivity parallel hole. (B) Converging. (C) All purpose parallel hole. (D) Diverging. (E) High resolution parallel hole. (F) Pinhole. (From Rollo FD, Harris CC. Factors affecting image formation. In: Rollo FD, ed. *Nuclear Medicine Physics, Instrumentation and Agents.* St. Louis: Mosby; 1977:407. Modified from Moyer RA. J Nucl Med 1974, 15:59.)

that collimator efficiency, E_g, for parallel-hole collimators is not affected by the source-to-detector distance for an extended planar source; that is, it remains the same at different distances from the detector. Collimator efficiency varies with different types of collimators, and the values are shown as a function of source-to-collimator distance in Figure 10.8.

Uniformity

It is always expected that an imaging device, primarily a gamma camera, should yield a uniform response throughout the field of view. That is, a point source counted at different locations in the field of view should give the same count rate by the detector at all locations. However, even properly tuned and adjusted gamma cameras produce nonuniform images with count density variations of as much as 10%. Such nonuniformity adds to the degradation of the spatial resolution of the system.

The nonuniformity in detector response primarily arises from the varia-

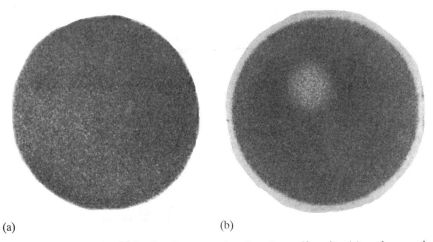

(a)   (b)

Fig. 10.9. Images of a ^{57}Co flood source showing the uniformity (a) and nonuniformity (b) of the response of a gamma camera.

tions in PM tube response, that is, from variations in the production of pulses and from nonlinearities in X, Y positioning of the identical pulses across the field of view. Nonuniformity is also seen around the edge of an image as a bright ring, and it is called *edge packing*. This results from the fact that more light photons are reflected from the reflector material near the edge of the detector to the PM tubes. Normally a lead ring is attached to the edge of the collimator to mask this effect. Nonuniformity may also arise from the improper positioning of the photopeak by PHA window settings.

Uniformity in response of a gamma camera is checked by obtaining an image of a flood source (e.g., ^{57}Co) using the camera with the collimator on (Fig. 10.9). Visual inspection of the image should reveal any nonuniformity that warrants the tuning of the camera, particularly the PM tubes. Nowadays, uniformity has improved considerably with the better attachment of the PM tubes to the detector.

Even with well-tuned gamma cameras, there always exists some degree of nonuniformity in the images, and several methods have been developed to correct for these nonuniformities. Of all methods, the use of a microprocessor built into a gamma camera has been the common practice. Typically, a ^{99m}Tc source containing ~ 250 μCi (9.3 MBq) is counted at a distance of 1–2 m from the gamma camera without the collimator, and a test image is stored in a computer matrix. The microprocessor then generates correction factors for each pixel of the matrix, based on variations in counts in different pixels. In subsequent images of patients, these correction factors are applied to each pixel data, either to add counts to the "cold" areas or to subtract counts from the "hot" areas. By this technique, the nonuniformity can be reduced to $\pm 2\%$–3%.

Contrast

Contrast of an image is the relative variations in count densities between adjacent areas in the image of an object. Contrast (C) gives a measure of detectability of an abnormality relative to normal tissue and is expressed as

$$C = \frac{A - B}{A} \qquad (10.10)$$

where A and B are the count densities recorded over the normal and abnormal tissues, respectively.

Lesions on the image are seen as either "hot" or "cold" spots indicating increased or decreased uptakes of radioactivity in the corresponding areas in the object. Several factors affect the contrast of the image, namely, count density, scattered radiation, pulse pileup, type of film, size of the lesion, and patient motion, and each contributes to the contrast to a varying degree.

For a given imaging setting, a minimum number of counts need to be collected for reasonable image contrast. Even with adequate spatial resolution from the imaging device, lack of sufficient counts may give rise to poor contrast, so much so that lesions may be missed. The optimum count density is about 1000 counts/cm^2. This count density depends on the amount of activity administered and the uptake in the organ of interest. Contrast is improved with increasing administered activity and also with the differential uptake between the normal and abnormal tissues. However, due consideration should be given to the radiation dose to the patient from a large amount of administered activity. Sometimes, high count density is achieved by counting for a longer period of time in the case of low administered activity.

Background (noise) in the image increases with scattered radiations and thus degrades the image contrast. Narrow PHA window settings can reduce the scatter radiations, but sensitivity, that is, counting efficiency, is reduced by narrow window settings. A 20% PHA window centered on the photopeak of interest is most commonly used in routine imaging.

At high count rates, pulse pileup can degrade the image contrast. Two Compton events occuring simultaneously may add up to form the photopeak, but the event will be mispositioned somewhere between the two events and hence the distortion of the image.

Film contrast is a component of overall image contrast and depends on the type of film used. The density response characteristics of x-ray films are superior to those of Polaroid films and provide the greatest film contrast, thus adding to the overall contrast. Developing and processing of exposed films may add artifacts to the image and therefore should be carried out carefully.

Image contrast to distinguish a lesion depends on its size relative to system resolution and its surrounding background. Unless a minimum size of a lesion larger than system resolution develops, contrast may not be sufficient to appreciate the lesion, even at higher count density. The lesion size factor depends on the background activity surrounding it and on whether it is a "cold" or "hot" lesion. A relatively small-size "hot" lesion can be well contrasted against a lower background, whereas a "cold" lesion may be missed against surrounding tissues of increased activities.

Patient motion during imaging reduces the image contrast. This primarily results from the overlapping of normal and abnormal areas by the movement of the organ. It is somewhat alleviated by restraining the patients or by having them in a comfortable position.

Quality Control Tests for Gamma Cameras

To ensure a high quality of images produced by imaging devices, several quality control tests must be performed routinely on these devices. Because the gamma camera is the most common imaging device, quality control tests for this unit will be described. The frequency of tests is either daily or weekly. The most common tests are the positioning of the photopeak (peaking), the uniformity check and the spatial resolution of the camera. These tests are carried out with the collimator attached (extrinsic) or without the collimator (intrinsic). Because ^{99m}Tc is easily available and employed in most nuclear medicine studies, it is often used for these tests. On the other hand, ^{57}Co with a longer half-life emits photons of 122 and 136 keV, which are equivalent to the 140-keV photons of ^{99m}Tc, and is used in the form of sealed sheet sources. Whether to use ^{99m}Tc or ^{57}Co and the intrinsic or extrinsic mode of testing is a matter of choice by the individual user.

Daily Checks

Positioning of Photopeak

This test is done daily or as needed to center the PHA window on the photopeak of interest (peaking) by adjusting the baseline and window settings of the PHA of the camera. For ^{99m}Tc, normally 1 mCi (370 MBq) of the isotope in a syringe is placed on the collimator attached to the detector (extrinsic). The window settings are adjusted to set the window on the 140-keV peak. The peaking must be carried out for each photon energy that is to be used for the imaging study. Thus, peaking for ^{111}In, ^{67}Ga, ^{123}I, ^{201}Tl, and so on must be done separately.

Uniformity

The uniformity of the detector response is tested daily by using a flood source containing approximately 10 mCi (370 MBq) of ^{57}Co. The source is placed on the detector head with the collimator attached (extrinsic). An image is taken on x-ray film with sufficient counts and then assessed for uniformity by visual inspection. Nonuniformity exceeding $\pm 15\%$ or so only will be detected by the human eye. If nonuniformity is noted, it is commonly related to the problems in the PM tube response and improper positioning of the photopeak.

Nonuniformity is corrected by a microprocessor built into the gamma camera. An ~ 250-μCi (9.3-MBq) source of ^{99m}Tc is placed ~ 1–2 m away from the detector with the collimator removed (intrinsic), and a test image is stored in a matrix in the computer. The microprocessor generates correction factors for each pixel from these image data, which are then applied to subsequent images. Normally, the test image is taken once a week and used for all studies for the entire week.

Weekly Checks

Spatial Resolution

The spatial resolution of the gamma camera is checked weekly by using bar phantoms. The bar phantom is placed on the detector head with the collimator attached (extrinsic), and a flood source of ~ 10 mCi (370 MBq) of ^{57}Co is placed on the top of the bar phantom. An image is taken on x-ray film with sufficient counts and visually inspected to check the linearity and separation of the smallest bars.

In addition to the preceding routine quality-control tests, other tests on accessories such as computers, multiformat cameras, scanning tables, rotation of gantry, and so on should be performed periodically. Furthermore, all tests must be documented in a record book with pertinent information, such as the date, time, total counts, window settings, the type of radioactive source, the type of camera, and initials of the technologist performing the test.

Questions

1. (a) Define the spatial resolution of a gamma camera.
   (b) What are the different components of the spatial resolution?
   (c) A system with a spatial resolution of 5 mm is better than a system with a spatial resolution of 8 mm. True or False?
2. The intrinsic resolution of a gamma camera depends on:
   (a) The thickness of the NaI(Tl) detector. True or False?
   (b) The energy of the γ-ray. True or False?
   (c) The width of the pulse-height window. True or False?
   (d) The number of counts collected. True or False?
3. (a) What is the best photon energy for imaging with a gamma camera?
   (b) Why is a thinner NaI(Tl) detector used in a gamma camera?
   (c) Intrinsic resolution improves with higher γ-ray energy. True or False?
   (d) Spatial resolution of a gamma camera improves as the number of photomultiplier tubes is increased. True or False?

4. For a gamma camera with a parallel-hole collimator,
   (a) The spatial resolution increases with decreasing detector thickness. True or False?
   (b) The collimator efficiency decreases with increasing collimator length. True or False?
   (c) The spatial resolution increases with decreasing collimator length. True or False?
   (d) High-energy collimators have higher efficiency and resolution than low-energy collimators. True or False?
   (e) The best resolution is obtained at the face of the parallel-hole collimator. True or False?
5. What are the effects of the following factors on the spatial resolution and sensitivity of a gamma camera?
   (a) Photomultiplier (PM) tubes with higher quantum efficiency
   (b) A wider "window" on the pulse-height analyzer (PHA)
   (c) Increasing the activity of ^{99m}Tc from 5 mCi (185 MBq) to 15 mCi (555 MBq)
   (d) Increasing the diameter of the collimator hole
   (e) Adding more tissue between the collimator face and the patient's organ
   (f) Using a diverging collimator
   (g) Increasing the source-to-collimator distance for a parallel-hole collimator
   (h) Using a γ-ray of higher energy, which penetrates the septum of the collimator
6. (a) In routine practice, how is the spatial resolution of a gamma camera checked?
   (b) The full width at half maximum of the line spread function of a gamma camera does not give a true picture of spatial resolution. Why?
   (c) What is the modulation transfer function (MTF) of a system?
   (d) A system gives the best spatial resolution when its MTF is equal to 1. True or False?
   (e) If PM tubes and the PHA of a gamma camera have MTFs of 0.5 and 0.7 at a certain spatial frequency, what is the overall MTF of the camera?
   (f) As the sensitivity of a gamma camera increases, its spatial resolution decreases. True or False?
   (g) The collimator efficiency of a parallel-hole collimator is not affected by the source-to-detector distance. True or False?
7. (a) What is the primary cause of nonuniformity in an image?
   (b) What is edge packing?
   (c) How is the nonuniformity in an image corrected?
8. (a) What is the contrast of an image?
   (b) What are the different factors that affect the contrast of an image?
   (c) How does pulse pileup affect the contrast?

(d) Is contrast or spatial resolution affected by increasing the administered activity?
9. (a) What are the daily and weekly tests performed for gamma cameras?
   (b) What is meant by extrinsic and intrinsic tests?

References and Suggested Readings

Erickson J. Imaging systems. In: Harbert J, da Rocha AFG, ed. *Textbook of Nuclear Medicine. Volume I. Basic Science.* Philadelphia: Lea & Febiger; 1984:105.

Rollo FD. Evaluating imaging devices. In: Rollo FD, ed. *Nuclear Physics, Instrumentation and Agents.* St. Louis: Mosby: 1977:436.

Rollo FD, Harris CC. Factors affecting image formation. In: Rollo FD, ed. *Nuclear Physics, Instrumentation and Agents.* St. Louis: Mosby; 1977:387.

Sorensen JA, Phelps ME. *Physics in Nuclear Medicine.* 2nd ed. New York: Grune & Stratton; 1987, Chapter 16.

CHAPTER 11

Tomographic Imaging Devices

General Considerations

Conventional gamma cameras and rectilinear scanners provide two-dimensional planar images of three-dimensional objects. Structural information in the third dimension, depth, is obscured by superimposition of all data along this direction. Although imaging of the object in different projections (posterior, anterior, lateral, and oblique) can give some information about the depth of a structure, precise assessment of the depth of a structure in an object is made by tomographic scanners. The prime objectives of these scanners are to display the images of the activity distribution in different sections of the object at different depths.

The initial approach to the tomographic technique was to focus the image of the object at the plane of interest and at the same time to blur the images of the objects not in plane. This is called the focal plane method, or longitudinal tomography. Focused collimators are used to achieve images of the object at the focal plane. Seven-pinhole collimators (each pinhole projecting a different focal length) and slant-hole collimators have been used in longitudinal tomography to provide longitudinal section images. Even though out-of-focus plane images are blurred, however, the in-focus plane image still is obscured by counts from the former. Furthermore, because the imaging device is not rotated around the patient, only a limited number of angular views are available, as dictated by the collimator holes.

An improvement over focal plane tomography has been made by introducing transverse tomography in which multiple views are obtained at many angles around the patient, and images are constructed by computers. This method is called the emission computed tomography (ECT), which is based on sophisticated but complex mathematical algorithms. ECT furnishes images at distinct focal planes (slices) of the object. Illustrations of four tomographic slices of the heart are presented in Figure 11.1.

In nuclear medicine, two types of ECT have been in practice based on the type of radionuclides used: single photon emission computed tomography (SPECT), which uses γ-emitting radionuclides such as ^{99m}Tc, ^{123}I, ^{67}Ga,

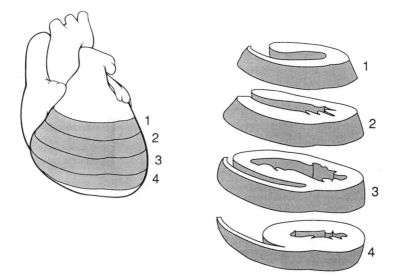

Fig. 11.1. Four slices of the heart in the short axis.

[111]In, and so forth, and positron emission tomography (PET), which uses β^+-emitting radionuclides such as ^{11}C, ^{13}N, ^{15}O, ^{18}F, ^{68}Ga, [82]Rb, and so forth. These two modalities of ECT are described below in detail.

Single Photon Emission Computed Tomography

The most common SPECT system consists of a typical gamma camera with one to three NaI(Tl) detector heads mounted on a gantry, an on-line computer for acquisition and processing of data, and a display system (Fig. 11.2). The detector head rotates around the long axis of the patient at small angle increments (3°–10°) for 180° or 360° angular sampling. The data are collected at each angular position and normally stored in a 64 × 64 or 128 × 128 matrix in the computer for later reconstruction of the images of the planes of interest. Transverse (short axis), sagittal (vertical long axis), and coronal (horizontal long axis) images can be generated from the collected data. Multihead gamma cameras collect data in several projections simultaneously and thus reduce the time of imaging. A three-head camera collects a set of data in about one third of the time required by a single-head camera.

Image Reconstruction

The most popular technique of image reconstruction from the collected data is based on the principles of so-called *backprojection*. The technique is illustrated in Figure 11.3, in which three projection views are obtained by a

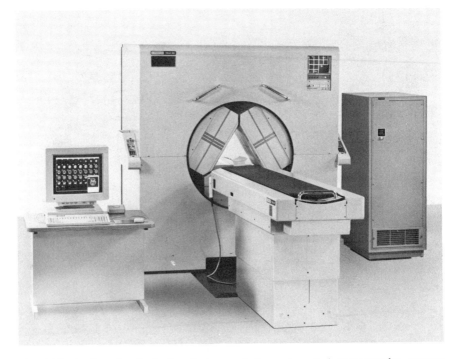

Fig. 11.2. A three-head single photon emission computed tomography camera, TRIAD model. (Courtesy of Trionix Research Laboratory, Inc., Twinsburg, Ohio.)

gamma camera at three equidistant angles around the object with two hot spots. The two-dimensional data-acquisition technique points to the fact that each single count in an image profile is the sum of all counts along a single line through the depth of the object (Fig. 11.3A). In the back projection technique, each count is then projected back along the line of collection perpendicular to the face of the detector (i.e., to backproject the data), whereby individual backprojected profiles superimpose to form an image (Fig. 11.3B). When many projection views are superimposed, a final image is obtained as shown in Figure 11.3C.

Backprojection can be better explained in terms of data acquisition in the computer matrix. Suppose the data are collected in a 4×4 acquisition matrix, as shown in Figure 11.4A. In this matrix, each row represents a slice of a certain thickness and is backprojected individually. Each row consists of 4 pixels. For example, the first row has pixels A_1, B_1, C_1, and D_1. Data in each pixel is considered to be the sum of all counts along the depth of the view. In the backprojection technique, a new reconstruction matrix of the same size (i.e., 4×4) is designed so that counts in pixel A_1 of the acquisition matrix are added to each pixel of the first column of the reconstruction matrix (Fig. 11.4B). Similarly, counts from pixels B_1, C_1, and D_1 are added

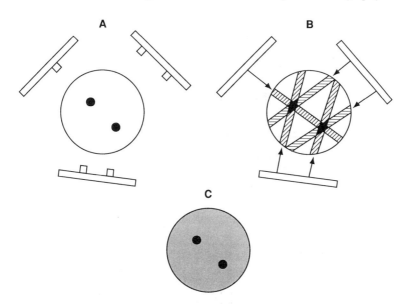

Fig. 11.3. Basic principle of reconstruction of an image by the backprojection technique. (A) An object with two "hot" spots (solid spheres) is viewed at three projections (at 120° angles). (B) Collected data are used to reconstruct the image by backprojection. (C) When many views are obtained, the reconstructed image represents the activity distribution with "hot" spots.

to each pixel of the second, third, and fourth columns of the reconstruction matrix, respectively.

Next, suppose a lateral view (90°) of the same object is taken, and the data are again stored in a 4 × 4 acquisition matrix. The first row of pixels (A_2, B_2, C_2, and D_2) in the 90° acquisition matrix is shown in Figure 11.4B. Counts from pixel A_2 are added to each pixel of the first row of the same reconstruction matrix, counts from pixel B_2 to the second row, counts from pixel C_2 to the third row, and so on. If more projection views are taken at angles between 0° and 90°, or any other angle greater than 90° and stored in 4 × 4 acquisition matrices, then counts from each pixel of the first row of every acquisition matrix can be added to each pixel of the corresponding diagonal row of the reconstruction matrix. This type of backprojection results in superimposition of data in each pixel, thereby forming the final transverse image with areas of increased or decreased activity (see Fig. 11.3C).

Using the 4 × 4 matrix, four transverse cross-sectional images (slices) can be produced. Similarly, using 64 × 64 matrices for both acquisition and reconstruction, 64 transverse slices can be generated. From all transverse slices, appropriate pixels are sorted out along the horizontal and vertical long axes and used to form coronal and sagittal images. It is a common practice to lump several slices together to increase the count density on the individual slices for better contrast.

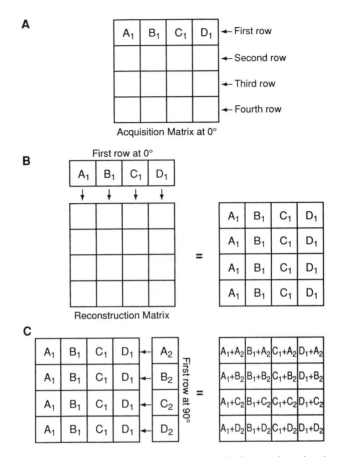

Fig. 11.4. An illustration of the backprojection technique using the data from an acquisition matrix into a reconstruction matrix.

The backprojection technique has the problem of the "star effect" artifacts caused by "shining through" radiations from adjacent areas of increased radioactivity (see Fig. 11.3). It is normally corrected by applying a "filter" to each individual row or column of the acquisition matrix, and then the filtered profiles are backprojected onto the reconstruction matrix to produce a sharper image without the star effect. Filtering is accomplished by assigning small negative values to each side of a peak (positive value). The peak may encompass one or more pixels of the acquisition matrix. These negative values will in effect cancel or erase all but the highest count areas of the rows or columns. This process is also called the *convolution method.* Various types of filters are commercially available in the form of software packages, most of which use the so-called *ramp filters.* The use of ramp filters in backprojection suppresses the star effect.

An alternative method of reconstructing the image from the acquired data is to apply the analytic technique using the Fourier transform. This method converts data from spatial domain into frequency domain and requires rigorous mathematical treatment, which is beyond the scope of this book. The convolution or filtered backprojection technique, however, still remains the most common method of image reconstruction in nuclear medicine.

Factors Affecting SPECT

Photon Attenuation

γ-ray photons are attenuated by body tissue while passing through a patient. Attenuation causes distortion in count density on the image. The degree of attenuation depends on the photon energy, the thickness of tissue and the linear attenuation coefficient of the tissue for the photon. Several techniques are employed to correct for attenuation. In one method, an uncorrected image is taken, and the thickness of tissue is estimated. Using a constant linear attenuation coefficient for the tissue, each pixel data is corrected to reconstruct the image. In another technique, a phantom simulating the organ of interest is used to generate correction factors, which are then applied to data acquired in real imaging of an object.

Assumption of a constant linear attenuation coefficient is not valid for several organs such as the heart, because of the close proximity to other organs. γ-rays traversing through different thicknesses of body tissue may be detected within the photopeak, and therefore a constant correction factor may not be sufficient for attenuation correction. Most often, attenuation corrections are not applied to SPECT images for reasons of complexity of the problem.

Center of Rotation

The center of rotation (COR) parameter is a measure of the alignment of the opposite views (e.g., posterior versus anterior or right lateral versus left lateral) obtained by the SPECT system. In other words, the COR must be accurately aligned with the center of the acquisition matrix in the computer. If the COR is misaligned, then a point source would be seen as a "donut" on the image (Fig. 11.5). Thus, an incorrect COR in a SPECT system would result in image degradation. For example, an error of 3 mm in the alignment of COR is likely to cause a loss of resolution of $\sim 30\%$ in a typical SPECT system.

The misalignment of COR may arise from improper shifting in camera tuning, mechanics of the rotating gantry, and misaligned attachment of the collimator to the detector. It is essential that the COR alignment is assessed routinely in SPECT systems (e.g., weekly) to avoid degradation in image resolution. Manufacturers provide detailed methods of determining COR alignment for SPECT systems, which should be included in the routine quality control procedure. The COR off by more than 1 pixel may cause degradation in the reconstructed image.

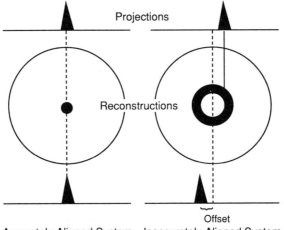

Fig. 11.5. An illustration of the effect of a misaligned center of rotation. A "donut"-shaped image appears from inaccurately aligned center of rotation. (From Todd-Pokropek A. The mathematics and physics of emission computerized tomography (ECT). In: Esser PD, Westerman BR, eds. *Emission Computed Tomography.* New York: Society of Nuclear Medicine 1983; 3.)

Sampling

It is understood that the larger the number of projections (i.e., small angular increment), the less is the star effect, and hence the quality of image is better. However, this requires a longer time of acquisition. On the other hand, a fewer number of projections may not erase the star effect. Therefore, one should choose an optimal number of angular sampling for appropriate image contrast. Angular sampling at 6° intervals in 360° (60 projections) is adequate for most SPECT systems.

Spatial Resolution

Spatial resolution of a SPECT system is affected by all factors that have been considered for the conventional camera, and therefore the readers should refer to Chapter 10. Typical SPECT systems have spatial resolution of 15–20 mm. Spatial resolution deteriorates, but sensitivity increases with increasing slice thickness. As a trade-off between resolution and sensitivity, an optimum slice thickness should be chosen.

Noise

Noise arises from background radiations, scatter radiations, or statistical variations in counts and may be present in filtered backprojected images. Overlapping of pixel counts along the backprojection line and low statistics of count rates add to the noise in SPECT images. Roughly, increasing the total acquisition counts decreases the effect of noise on the image.

Sensitivity

The sensitivity of an imaging system is always desired to be higher for better image contrast. The SPECT systems are designed for greater sensitivity so that high counts can be accumulated in a reasonable time for images of thin slices of an organ. For conventional two-dimensional planar images of good contrast, about 500,000 counts are required. Thus, if each sectional image (i.e., slice) of an organ requires 500,000 counts for the same contrast as in a conventional image, and if there are, for example, 20 sectional images of an organ of interest, then 10 million counts would be needed for the entire organ. For most SPECT systems using low-energy all-purpose collimators, 5–20 million counts are needed. Total counts may be increased either by counting for a longer period or by administering more activity.

Positron Emission Tomography

PET is based on the detection in coincidence of the two 511-keV annihilation radiations that are emitted by the β^+-emitting radionuclides from a source (e.g., the patient). Positrons are annihilated in body tissue and produce two 511-keV annihilation photons that are emitted at 180°. Two photons are detected by two detectors connected in coincidence, and data collected over 360° around the body axis of the patient are used to reconstruct the image of the activity distribution in the slice of interest. A schematic diagram of the PET system using four pairs of detectors is illustrated in Figure 11.6.

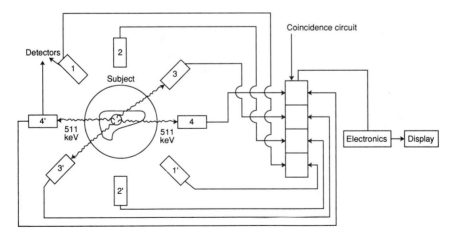

Fig. 11.6. A schematic diagram of a positron emission tomographic imager with four pairs of detectors.

Table 11.1. Characteristics of PET radionuclides.

Radionuclides	Half-life	Modes of decay (%)	$\bar{E}_{\beta^+}$ (MeV)
^{11}C	20.4 min	β^+ (100)	0.385
^{13}N	10 min	β^+ (100)	0.491
^{15}O	2 min	β^+ (100)	0.735
^{18}F	110 min	β^+ (97)	0.242
		EC (3)	
^{68}Ga	68 min	β^+ (89)	0.740
		EC (11)	
^{82}Rb	75 sec	β^+ (95)	1.409
		EC (5)	

The detectors are primarily made of bismuth germanate (BGO), NaI(Tl) or barium fluoride (BaF$_2$), of which BGO is most commonly used in the current PET systems. BGO is not hygroscopic, so hermetic sealing is not required. The scintillation decay time of BGO is somewhat longer than those of NaI(Tl) and BaF$_2$.

In current PET systems, many detectors (hundreds to thousands) are arranged in circular, hexagonal, or octagonal rings. The number of rings can vary from 1 to 16, and they are arranged in arrays. For example, the Posicam 6.5 camera (Positron Corporation, Houston, Tex.) uses 1320 BGO detectors arranged in an array of 11 rings with 120 detectors in each ring. Each detector is connected to the opposite detector by a coincidence circuit. The field of view is defined by the width of the array of detectors, and the source-to-detector distance is defined by the radius of the array. Because photons are detected in coincidence only along a straight line, no collimator is needed to limit the field of view. This coincidence technique is often termed the *electronic collimation*.

Coincident counts from the entire field of view over 360° angles around the patient are acquired simultaneously in a 64 × 64, 128 × 128, or 256 × 256 matrix in a computer. Data can be collected in either the frame mode or the list mode. Reconstruction of images is performed by backprojection after filter correction of the acquired data in the same way as in the SPECT system.

The most common radionuclides for PET imaging are listed in Table 11.1. Of these, ^{82}Rb in the form of RbCl is used for myocardial perfusion imaging. ^{18}F-fluorodeoxyglucose (FDG) is primarily used for metabolic imaging in the heart and brain.

Factors Affecting PET

Attenuation

It is understandable that the two 511-keV annihilation photons originate from all regions of an organ of interest, and therefore off-center photons

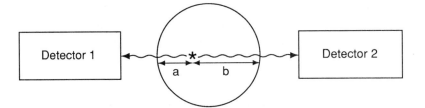

Fig. 11.7. Two 511-keV annihilation photons being attenuated by different thicknesses of tissue.

traverse different thicknesses of tissues to reach the two opposite detectors (Fig. 11.7). These off-center photons will be attenuated to different extents, and attenuation corrections should be made to the counts before reconstruction of images. It can be shown analytically that the attenuation factor does not depend on the location of the origin of the photons but rather on the total tissue thickness of the organ.

There are two methods for attenuation corrections (Hoffman and Phelps, 1986): theoretical method and transmission scan method. In the theoretical method, attenuation correction factors are estimated from the knowledge of linear attenuation coefficient μ and the assumption of the contour of the organ of interest. Inaccuracies in the assumption of the body contour result in uncertainty in these correction factors. This method is good for organs such as the brain, which are uniform in density and not surrounded by other organs.

In the transmission scan method, which gives a direct measure of attenuation, a plexiglass circular ring containing positron emitters (e.g., ^{68}Ga) is placed between the detector ring and the body of the patient, keeping the organ in the field of view. Alternatively, a thin rod source containing a long-lived positron emitter (e.g., ^{68}Ge → ^{68}Ga; ^{68}Ge has a half-life of 270 days) is rotated around the patient by a motor. Two transmission scans are obtained, one with the patient in place and one without the patient. Correction factors are obtained from the ratio of the two counts for each pixel and then applied to the data in the corresponding pixel of the reconstruction matrix. In practice, the transmission scan without the patient is not obtained for every patient and instead is acquired periodically (usually weekly), which is then used to calculate the correction factors for each patient in that period. Although this method is more accurate than the theoretical method, it has an inherent error arising from statistical noise in the transmission data.

Random and Scatter Coincidences

Random or accidental coincidence events are those annihilation photons that are not in coincidence but accidentally counted within the coincidence timing. Random events may arise from annihilation radiations from inside and outside the field of view of the detectors or from regular γ-rays of similar energy (i.e., 511 keV) from the same β^+ emitters or other radionuclides. Random coincidences normally overestimate the count density in the center of the image and should be corrected (Hoffman and Phelps, 1986). Random events can be reduced by using a shorter coincidence timing and increasing the radius of the ring of detectors.

Random coincidence corrections are made by measuring two single count rates (R_1 and R_2) by the two detectors and by using the equation

$$R_c = 2\tau R_1 R_2 \tag{11.1}$$

where R_c is the random coincidence count rate and τ is the coincidence timing, which is set by the user for the system. In another method, one signal is delayed well beyond the coincidence timing relative to the other signal, and therefore no true coincidences are recorded. Only random coincidences, if any, will be recorded. In either case, random counts are then subtracted from the total coincidence counts.

Annihilation radiations may undergo scattering while passing through the body tissue, and most of these scattered radiations move in the forward direction without much loss of energy (Hoffman and Phelps, 1986). These scattered radiations are energetically indistinguishable from the true coincidences and therefore may be counted as true coincidences. These radiations can originate from inside and outside the field of view but still be detected by the detector. But the images are obscured by uncertainty in the locations of scattered radiations. The scatter contribution depends on the depth of the body tissue, the activity distribution in different cross-sections of the object, and the electronics of the PET system. Practically, the correction for scatter is made by measuring the activity at the edge of the field of view (where no true coincidence counts are expected) and subtracting this value from all the recorded data across the field of view.

Spatial Resolution

Spatial resolution is determined mainly by detector size and improves with smaller detectors. The ring diameter and the detector material also affect the intrinsic resolution of the PET system. PET systems are designed to achieve a specific spatial resolution, thus dictating the different dimensions of the detector.

Positrons traverse a certain distance in tissue, losing most of their energy before annihilation and then are annihilated away from the site of their origin. Thus, the site of β^+ emission differs from the site of annihilation photons. The distance of positron path increases with the positron energy. This phenomenon degrades the spatial resolution. Therefore, high-energy β^+ emitters give poorer spatial resolution than low-energy β^+ emitters. For example, myocardial perfusion imaging with $^{13}NH_3$ ($\bar{E}_{\beta^+} = 0.491$ MeV) will give better spatial resolution than that with ^{82}Rb ($\bar{E}_{\beta^+} = 1.409$ MeV). Furthermore, all annihilation photons are not always emitted at 180°, because the positron-electron pair is not completely at rest during annihilation. Such angular deviations may be of the order of 0.25° and can degrade the spatial resolution.

The overall spatial resolution of current PET systems is about 4–8 mm.

Sensitivity

The sensitivity of PET systems depends on the detector material, slice thickness, and the diameter of the ring. Sensitivity increases with slice thickness at the loss of resolution along the depth. It also increases with decreasing ring diameter as a result of increasing number of coincidence events detected.

Quality Control Tests for SPECT and PET

Daily

Uniformity

The uniformity test for the SPECT camera is the same as that for the conventional gamma camera described in Chapter 10.

In the case of the PET camera, uniformity can be checked by imaging a large elliptical or circular phantom filled with a positron emitter (e.g. ^{68}Ga). The variations in counts of different regions of each tomographic slice are calculated by a computer and the uniformity of the image is assessed from these variations.

Weekly

Spatial Resolution

For single-head and dual-head SPECT systems, spatial resolution is checked by using bar phantoms in the same manner as for conventional gamma cameras. For triple-head SPECT cameras (e.g., TRIAD), the manufacturer's service personnel check for spatial resolution periodically.

For PET systems, the manufacturer's service personnel check for the spatial resolution periodically using line spread functions and therefore no specific test is warranted routinely.

Center of Rotation

For single photon emission computed tomography (SPECT) systems, the alignment for the center of rotation is checked weekly. Commercial software packages are available for this test. A small spot marker of ^{57}Co or any other γ-emitting radionuclide (~ 1 mCi, i.e., 37 MBq) is placed at a distance from the detector with the parallel-hole collimator attached. The position of the source is off-centered; that is, at some distance away from the central axis of the detector. Approximately 20,000 counts are acquired at appropriate window settings at each angular projection and stored in a matrix of choice, for example, a 64 × 64 matrix, in the computer. Data are collected over 360° for single-head gamma cameras and over 180° for dual-head cameras. The computer program then computes the positions of the source along the X and Y axes in the computer matrix as seen by the one (360°) or two (180°) detectors. These positions are expressed in terms of pixel numbers, which are then plotted against the angles of rotation. The X-axis plot should give a symmetrical bell-shaped curve, and the Y-axis plot should be a straight line passing through the pixel on which the source is positioned, indicating the proper alignment of the center of the acquisition matrix with the center of the detector. The COR can also be calculated from the sum of the pixel numbers of the exactly opposite views divided by two. A deviation of more than one pixel warrants the servicing of the camera.

Questions

1. Describe the principles of single photon emission computed tomography (SPECT).
2. Explain how SPECT images are reconstructed by the backprojection technique.
3. Elucidate different factors that affect SPECT images.
4. What is the optimum angular sampling for the SPECT system?
5. How does the center of rotation affect the SPECT image, and what does it originate from?
6. Describe the principles of positron emission tomography (PET).
7. What type of detector is most commonly used in PET?
8. How is the attenuation of annihilation photons corrected for in PET?
9. How do random and scatter radiations affect PET images? What are the methods for correcting these effects?
10. What are the typical values of spatial resolutions of SPECT and PET systems?
11. The "star" effect can be reduced by increasing the number of projections around the object of study. True or False?
12. Filters are used in backprojection to cancel or lose all but the highest density in the image. True or False?
13. What is the effect of ramp filters on the reconstructed image?
14. What are the most common radionuclides used in PET?
15. Explain why ^{13}N-ammonia gives better spatial resolution than ^{82}Rb in the PET imaging of myocardium.
16. What are the factors that affect the spatial resolution of PET?
17. Describe the method for determining the alignment of the center of rotation of a SPECT system.

References and Suggested Readings

Croft BY. *Single Photon Emission Computed Tomography*. Chicago: Year Book Medical Publishers; 1986.

English RT, Brown SE. *Single Photon Emission Computed Tomography: A Primer*. 2nd ed. New York: Society of Nuclear Medicine; 1990.

Hoffman EJ, Phelps ME. Positron emission tomography: principles and quantitation. In: Phelps ME, Mazziotta J, Schelbert H, eds. *Positron Emission Tomography and Autoradiography: Principles and Applications for the Brain and Heart*. New York: Raven Press; 1986:237–286.

Koeppe RA, Hutchins GD. Instrumentation for positron emission tomography: tomographs and data processing and display systems. *Semin Nucl Med.* 1992; XXII:162–181.

Phelps ME. Emission computed tomography. *Semin Nucl Med.* 1977;7:337–365.

Rollo FD, ed. *Nuclear Medicine Physics, Instrumentation, and Agents*. St Louis: CV Mosby; 1977.

Sorensen JA, Phelps ME. *Physics in Nuclear Medicine*. 2nd ed. New York: Grune & Stratton; 1987.

CHAPTER 12

Radiation Biology

The subject of radiation biology deals with the effects of ionizing radiations on living systems. During the passage through living matter, radiation loses energy by interaction with atoms and molecules of the matter, thereby causing ionization and excitation. The ultimate effect is the alteration of the living cells. The following section is a brief outline on radiation biology, highlighting the mechanism of radiation damage and different types of its effects on living matter.

The Cell

The cell is the building unit of living matter and consists of two primary components: the nucleus and the cytoplasm (Fig. 12.1). All metabolic activities are carried out in the cytoplasm under the guidance of the nucleus.

The nucleus contains chromosomes, which are composed of genes. Genes are made of deoxyribonucleic acid (DNA), which are responsible for conducting metabolic activities and propagating hereditary information in the cell. The sequence of genes in the chromosome characterizes a particular chromosome. Two categories of cells—namely, germ cells (reproductive cells such as oocytes and spermatozoa) and somatic cells (all other cells)—are based on the number of chromosomes they contain. Whereas germ cells contain n number of individual chromosomes, somatic cells contain $2n$ number of chromosomes in pairs, where n varies with species of the animal. In humans, n is equal to 23; therefore, there are 23 chromosomes in germ cells and 46 chromosomes in somatic cells.

In the cytoplasm of the cell exist several important organelles, which carry out the cellular metabolic activities. Ribosomes are made up of protein and ribonucleic acid (RNA) and are responsible for protein synthesis in living matter. Endoplasmic reticula are tubular structures mostly responsible for protein synthesis. Mitochondria are ellipsoidal structures with a central cavity and contain specific enzymes to oxidize carbohydrate and lipid to pro-

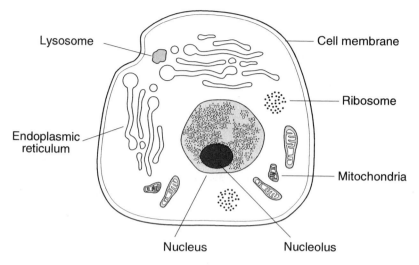

Fig. 12.1. Structure of a typical mammalian cell.

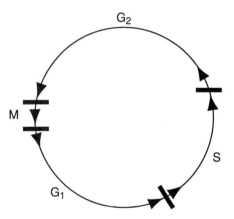

Fig. 12.2. The cell cycle. S is the DNA synthesis phase. M is the period of mitosis during which the prophase, metaphase, anaphase, and telophase take place. G_1 is the period between the telophase and S, and G_2 is the period between S and the prophase.

duce energy. Lysosomes are small organelles in the cytoplasm that contain enzymes capable of lysing many nutrients and cells.

The entire cytoplasm is enclosed within a cell membrane made of lipids and proteins. Its primary function is to selectively prohibit or permit the passage of substances into and out of the cell.

The growth of living matter is caused by proliferation of cells by cell division—a process in which a cell divides into two cells. The cell division of somatic cells is called *mitosis* and that of germ cells is called *meiosis*. Both mitosis and meiosis consist of four phases: prophase, metaphase, anaphase, and telophase. Each of these phases involves the rearrangement of the number of chromosomes and represents the progression of cell division.

Before cell division, each cell undergoes a long period, termed *interphase*, when DNA molecules are synthesized. In DNA synthesis, two new DNA molecules are produced, which are exact replicas of the original DNA molecule. This period of DNA synthesis is designated the "S" phase, which takes place around the middle of the interphase. The period between the telophase and the S phase is termed G_1 and the period between the S phase and the prophase is termed G_2 (Fig. 12.2). During the G_1 and G_2 periods, no functional activity related to cell division occurs. Of all phases, G_1 is the most variable phase in the cell cycle.

One important difference between mitosis and meiosis is that in meiosis, for a given series of cell division, every alternate cell division skips DNA synthesis, thus keeping the number of chromosomes the same in germ cells.

Radiation Units

Three units of measure are related to radiation: the *roentgen* (R) for exposure, the *rad* (radiation absorbed dose) for absorbed dose, and the *rem* (roentgen equivalent man) for dose equivalent.

The roentgen is the amount of x- or γ-radiation that produces ionization of one electrostatic unit of either positive or negative charge per cubic centimeter of air at 0°C and 760 mm Hg, standard temperature and pressure (STP). Because 1 cm^3 air weighs 0.001293 g at STP and a charge of either sign carries 1.6×10^{-19} C or 4.8×10^{-10} electrostatic units, it can be shown that

$$1 \text{ R} = 2.58 \times 10^{-4} \text{ C/kg} \tag{12.1}$$

It should be noted that the roentgen applies only to air and to x- or γ-radiations. Because of practical limitations of the measuring instruments, the R unit is applicable only to photons of less than 3 MeV energy.

The rad is a more universal unit. It is a measure of the energy deposited per unit mass of any material by any type of radiation. The rad is specifically defined as

$$1 \text{ rad} = 100 \text{ ergs/g absorber} \tag{12.2}$$

Because 1 joule (J) $= 10^7$ ergs,

$$1 \text{ rad} = 10^{-2} \text{ J/kg} \tag{12.3}$$

In System Internationale (SI) units, the *gray* (Gy) is the unit of radiation absorbed dose and is given by

$$1 \text{ Gy} = 100 \text{ rad} \tag{12.4}$$

$$= 1 \text{ J/kg absorber} \tag{12.5}$$

It can be shown that the energy absorbed per kilogram of air from an exposure of 1 R is

$$1 \text{ R} = 86.9 = 10^{-4} \text{ J/kg in air}$$

Therefore,

$$1 \text{ R} = 0.869 \text{ rad in air}$$

or

$$1 \text{ R} = 0.00869 \text{ Gy in air}$$

The rad is not restricted by the type of radiation or absorber or by the intensity of the radiation. It should be understood that the rad is independent of the weight of the material. This means that a radiation dose of 1 rad (0.01 Gy) is always 1 rad (0.01 Gy) in 1, 2, or 10 g of the material. The integral absorbed dose, however, is given in units of gram-rad (g · rad or g · Gy) and is calculated by multiplying the rad (Gy) by the mass of material. For example, if the radiation dose to a body of 45 g is 10 rad (0.1 Gy), then the integral radiation dose to the material is 450 g · rad (or 4.5 g · Gy); however, the radiation dose is still 10 rad (0.1 Gy).

The dose-equivalent unit, rem, has been developed to account for the differences in effectiveness of different types of radiation in causing biological damage. In radiobiology, the rem is defined as

$$\text{rem} = \text{rad} \times \text{RBE} \tag{12.6}$$

where RBE is the relative biological effectiveness of the radiation. It is defined as the ratio of the dose of a standard radiation to produce a particular biological response to the dose of the radiation in question to produce the same biological response. Radiations of 250 kV x-rays are normally chosen as the standard radiation because of their widespread use.

In radiation protection, however, the rem is defined as

$$\text{rem} = \text{rad} \times QF \times N \tag{12.7}$$

where QF is the quality factor, and N is the modifying factor of the radiation in question. The factor N, for all practical purposes, is assumed to be unity. QF is related to the linear energy transfer (LET) of the radiation in a given medium (see Chapter 6) and reflects the effectiveness of the radiation to cause biological or chemical damage similar to RBE. It is particularly useful in the designing of shielding and in the calculation of radiation dose to radiation workers. The QF values of various types of radiation are listed in Table 12.1.

In SI units, the dose equivalent is expressed in *sievert*, which is defined as

$$1 \text{ sievert (Sv)} = 100 \text{ rem} \tag{12.8}$$

In practical situations, all these radiation units are often expressed in milliroentgens (mR), millirad (mrad), and millirem (mrem), which are 10^{-3} times the units, roentgen, rad, and rem, respectively. In SI units, the equivalent quantities are milligrays (mGy) and millisieverts (mSv). A rad is also commonly expressed as centigray (cGy), one-hundredth of a gray.

Table 12.1. Quality factors for different radiations.

Type of radiation	QF
X-rays, γ-rays, β particles	1.0
Neutrons and protons	10.0
α particles	20.0
Heavy ions	20.0

Effects of Radiation on Cells

The nucleus of the cell is the most sensitive part to radiation and can undergo severe changes upon interaction with ionizing radiations. This sensitivity has been mostly attributed to the DNA molecule, which is broken down by radiation in various parts of the molecule. Any change in DNA structure by ionizing radiations is called a *mutation*. The number of mutations in the DNA molecule increases with increasing radiation exposure.

Chromosomes are likely to be affected by mutations of the DNA molecules. Chromosomes may themselves be affected by radiation, causing single or double breaks in the structure of the chromosome. Such alterations are called *chromosome aberrations*. Both DNA mutations and chromosome aberrations can propagate through cell division to future cell generations or may be repaired.

Whether chromosome aberration is induced by single breaks or double breaks in the structure determines the fate of the cell. In single breaks, the chromosome tends to repair by joining the two fragments in a process called *restitution*, provided sufficient time is allowed for the two ends to meet. The cell becomes normal functionally and replicates normally. If the chromosome does not repair, the cell becomes defective and replicates abnormally, producing defective cells, but hardly dies. In single breaks, almost all cells tend to survive through repair by restitution. Single breaks are caused mainly by low doses of radiation and low-LET radiations.

When chromosome aberration is caused by double breaks in the structure, the chances of two fragments joining in a correct sequence of genes are low, and therefore normal repair becomes difficult. If repair occurs at all, the cell becomes defective and replicates abnormally. Double breaks are mostly caused by high dose rates and high-LET radiations.

Repair of chromosomes after irradiation depends on the sites of break in the DNA molecule or the chromosome, the total radiation dose, the dose rate, and the LET of the radiation. Chromosome aberrations by double breaks occur more frequently at high dose rates than at low dose rates because of less time to repair and less chances of combining two fragments in correct sequence of genes. High-LET radiations cause more double breaks in chromosomes than low-LET radiations and thus repair becomes difficult.

For example, α particles, protons, and neutrons will cause more radiation damage than γ-rays.

Direct and Indirect Action of Radiation

The DNA molecule of a cell is the most sensitive target to radiation. Radiation damage to the cell can be caused by the direct or indirect action of radiation on the DNA molecules. In the direct action, the radiation hits the DNA molecule directly, disrupting the molecular structure. The structural change leads to cell damage or even cell death. Damaged cells that survive may later induce carcinogenesis. This process becomes predominant with high-LET radiations such as α particles and neutrons.

In the indirect action, the radiation hits the water molecules, the major constituent of the cell, and other organic molecules in the cell, whereby free radicals such as perhydroxyl (HO_2) and alkoxy (RO_2) are produced. Free radicals are characterized by an unpaired electron in the structure which is very reactive and therefore react with DNA molecules to cause cellular damage. The number of free radicals produced by ionizing radiations depends on the total dose but not on the dose rate. The majority of radiation-induced damage results from the indirect action mechanism, because water constitutes nearly 70% of the cell composition.

Radiosensitivity of Cells

In living matter, there are two types of cells: differentiated and undifferentiated cells. Undifferentiated cells are those cells that do not have any specific physiologic function except developing into mature cells. Undifferentiated cells undergo mitosis and serve as the precursors for mature cells. In contrast, all mature cells are differentiated and perform specific functions in the living body. For example, red blood cells (RBCs) are mature and differentiated cells performing the function of oxygen carriers, whereas erythroblasts are undifferentiated cells that develop into RBCs through mitosis.

According to the law of Bergonié and Tribondeau, *undifferentiated cells that are undergoing active mitosis are most sensitive to radiation, and differentiated or mature cells are least affected by radiation.* For example, in a sample of mixed RBCs, erythroblasts are most damaged and RBCs are least affected by radiation. Undifferentiated cells that are killed by radiation may be replaced by new cells, but those that survive with defective DNAs can induce late effects, such as cancer (see later). In contrast, the S phase (see Fig. 12.2) of DNA synthesis in the cell cycle is least radiosensitive. Radiosensitivity is best assessed by cell death. For differentiated cells, it means loss of cellular function, whereas for undifferentiated cells, it means loss of reproductivity.

Groups of cells and their relative radiosensitivity are listed in Table 12.2.

Table 12.2. Radiosensitivity of different types of cells.

Types of cells	Radiosensitivity
Mature lymphocytes Erythroblasts Spermatogonia	Highly sensitive
Myelocytes Intestinal crypt cells Basal cells of epidermis	Relatively sensitive
Osteoblasts Spermatocytes Chondroblasts Endothelial cells	Intermediate sensitivity
Spermatozoa Granulocytes Erythrocytes Osteocytes	Relatively resistant
Nerve cells Muscle cells Fibrocytes	Highly resistant

Adapted from Casarett AP. *Radiation Biology*. Englewood Cliffs, NJ: Prentice-Hall, Inc; 1968:168–169.

As can be seen, lymphocytes are most sensitive to radiation, owing to a large nucleus; nuclear material is more radiosensitive. Nerve cells and muscle cells are totally differentiated cells and are, therefore, highly resistant to radiation. The tissue or organ that contains more radiosensitive cells will be highly radiosensitive and vice versa. For example, bone marrow containing radiosensitive erythroblasts is very radiosensitive, whereas nerves and muscles containing radioresistant cells are less radiosensitive.

Cell Survival Curves

When mammalian cells are irradiated, not all cells are affected to the same extent. Different factors such as the total dose, the dose rate, the LET of the radiation, the particular stage of the cell cycle (M, G_1, S, or G_2), and the type of cell will affect the radiation-induced damage. Some cells may die, and others will survive. The cellular response to radiation is illustrated by what is called the *cell survival curve*. It is obtained by plotting the dose along the linear X-axis and the surviving fraction along the logarithmic Y-axis. Surviving cells are those cells that retain all reproductive as well as functional activities after irradiation, whereas the death of cells is indicated by the loss of their function in differentiated cells and by the loss of reproductive activity in undifferentiated cells.

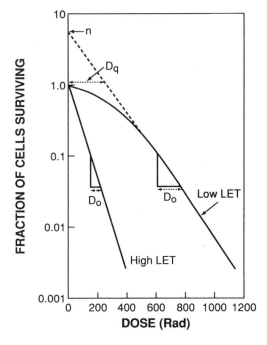

Fig. 12.3. Typical cell survival curves. The cell survival curve for low-linear energy transfer (LET) radiations shows a shoulder of width D_q, which is called the quasithreshold dose. After D_q, the plot becomes linear on a semilog scale, indicating an exponential dose–response relationship. The extrapolation number n is obtained by extrapolating the linear portion of the curve back to the ordinate. D_0 is the slope of the linear portion of the curve and is the dose at which 37% of the cells survive. The survival curve for high-LET radiations shows no or little shoulder, indicating D_q to be zero and n to be unity.

Typical cell survival curves are shown in Figure 12.3. For high-LET radiations such as α particles and low-energy neutrons, the survival curves are nearly straight lines starting from the lowest doses. In contrast, for low-LET radiations (e.g., x- and γ-radiations), the survival curve exhibits a shoulder initially followed by a straight line. This straight line portion on the semilog plot is an exponential curve on a linear plot. This curve is characterized by three parameters: D_0 (dose at which 37% of cells survive); the extrapolation number, n; and the quasithreshold dose, D_q.

The quasithreshold dose, D_q, is the dose at which the cell survival curve tends to become a straight line and is given by the width of the shoulder of the curve. The D_q indicates that, at low doses, almost all cells repair after irradiation and that cell killing is minimal, which is due to very limited radiation damage to the cell.

D_0 is determined from the slope of the straight line portion of the survival curve. It is the dose that kills 63% of the total number of cells. The value of

D_0 is a measure of radiosensitivity of a given type of cell. For example, a large value of D_0 for a type of cell means that the cells are less radiosensitive and vice versa.

The extrapolation number n is obtained by extrapolating the straight line portion of the survival curve back to the Y-axis. Its value depends on the width of the shoulder of the survival curve, that is, the quasithreshold dose, D_q. Its value for mammalian cells varies between 1 and 10.

Factors Affecting Cell Response to Radiation

As mentioned, various factors affect the radiation damage in the cell and hence the survival curve. The dose rate, the LET of the radiation, presence of chemical molecules, and the stage of the cell cycle all affect the survival curve.

Dose Rate

The dose rate, that is, the delivery of dose per unit time, is a prime factor in cellular damage. The higher the rate of dose delivery, the greater will be the cell damage. At low dose rates, only single breaks of chromosomes occur, and so cells have time to repair, whereas at high dose rates, double breaks occur, and so repair is less likely to occur because of the shorter time available to the cells between ionizing events. Figure 12.4 illustrates the effects of two dose

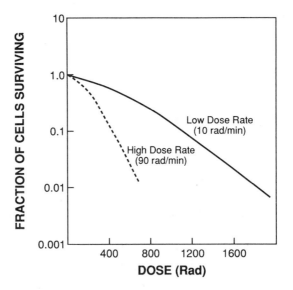

Fig. 12.4. The cell survival curves indicating the effect of dose rates. At high dose rates, the shoulder of the curve is reduced, with smaller values of D_q and larger values of n. The opposite is true at low dose rates.

rates on the cell survival curve. The dose rate effect is very important in radiation therapy, because unless an appropriate dose rate is prescribed, intended therapeutic effect may not be achieved. When a total dose is given to a patient in fractions over a period of time, it should be kept in mind that the interval between fractional doses should be short enough to keep repair of sublethal cellular damage to a minimum.

Linear Energy Transfer

High-LET radiations do not exhibit a dose-rate effect on the survival curve. Also at high dose rates (above 100 rad/min) of low- and moderate-LET radiations, no dose-rate effects are observed on the survival curve in contrast to low dose rates. Thus, high-LET radiations exhibit no shoulder (i.e., no D_q) on the survival curve, resulting in an extrapolation number of 1. High-LET radiations are densely ionizing radiations causing more double breaks in the chromosomes and thus causing more cell deaths than low-LET radiations, which are sparsely ionizing radiations. Radiation damage by high-LET radiations is so severe that the chances of repair are minimal, and even if repair takes place, the cell is likely to be defective.

Chemicals

Several chemicals, if present during the application of radiation, have been found to augment or diminish the effects of radiation on cells. Agents that enhance the cell response to radiation are called *radiosensitizers* and those that protect cells from radiation-induced damage are called *radioprotectors*.

Radiosensitizers

Oxygen is the most well-known sensitizer encountered in radiation biology. It has been found that hypoxic cells are very resistant to radiation, whereas oxygenated cells are highly radiosensitive. Such radiosensitization of cells by oxygen is called the *oxygen effect* and is measured by a quantity called the *oxygen enhancement ratio* (OER). The OER is given by the ratio of the dose required to produce a given radiation damage to cells in the absence of oxygen to that required to produce the same damage in the presence of oxygen. The oxygen effect occurs only when it is administered simultaneously with radiation. It occurs at O_2 tension up to 30 mm Hg, and no effect is observed at higher O_2 tension.

Figure 12.5 illustrates the effects of oxygen on the survival curve. The presence of oxygen makes the curve much steeper, indicating the augmentation of cellular damage at smaller doses relative to the situation of no oxygen. The mechanism of the oxygen effect is not clearly understood. It has, however, been postulated that oxygen combines with already formed free radicals, $R \cdot$, to produce peroxidyl group RO_2^-, which is more damaging to the cell components. Furthermore, while normally, $R \cdot$ could recombine with

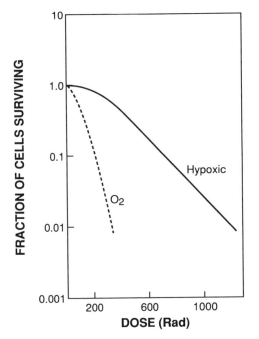

Fig. 12.5. The cell survival curve illustrating the effect of oxygen. In the presence of oxygen, the curve becomes steeper, indicating effective killing of the cells by radiation.

complimentary molecular components to repair the cell, RO_2 is an altered chemical entity and cannot help in cell repair. The oxygen effect is most predominant for γ- and x-rays and is practically absent for high-LET radiations (e.g., α particles).

Because tumor cells are mostly hypoxic, treatment of tumors under high-pressure oxygen has been advocated in conjunction with radiation. The OER value reaches a maximum of 3.0 for x- and γ-radiations, whereas it is about unity for high-LET radiations such as α particles.

Halogenated pyrimidines such as 5-chlorodeoxyuridine (ClUDR), 5-bromodeoxyuridine (BUDR), and 5-iododeoxyuridine (IUDR) are useful radiosensitizers. When cells are treated with these drugs for several days before irradiation with x- or γ-rays, cells become highly sensitive to radiation. Potentiation of radiosensitivity is due to the fact that these drugs are similar to the DNA precursor, thymidine, and are, therefore, incorporated into the DNA molecule, making them more susceptible to damage by radiation. For optimal therapeutic gain in radiotherapy, patients should be treated for a period of time covering several cell cycles to maximize drug incorporation into the cells.

Other radiosensitizers such as actinomycin D, puromycin, methotrexate, and 5-fluorouracil have been successfully used in combination with radiation to treat cancer. Whether these agents truly increase radiosensitivity or they are simply toxic to the cells is still not clear.

Investigators have been trying to explore radiosensitizing chemicals to substitute for oxygen, which requires the use of a high-pressure technique.

Metronidazole, (Flagyl) having a structure with high electron affinity, is a good radiosensitizer for hypoxic cells. Another very useful radiosensitizer for hypoxic cells is misonidazole, which also has high electron affinity. Misonidazole is almost 10 times more effective than metronidazole in sensitizing hypoxic cells.

Radioprotectors

The most common radioprotectors—substances that protect cells from radiation damage—include substances containing sulfhydryl groups, such as cysteine, cysteamine, and cystamine. Newer agents such as ethylphosphorathionic acid (WR-2721) and 2-mercaptopropionyl have been used with an increased protective factor and less toxicity. These agents protect cells from radiation damage by combining with free radicals. Radioprotectors are most effective for γ- and x-radiations and least effective for α radiations, because radiation damage is more drastic in the latter case.

Stage of Cell Cycle

Radiation damage mostly occurs during the period of mitosis, the M phase, whereas least damage occurs during the DNA synthesis, the S phase. Thus, the stage of the cell cycle determines the extent of radiation damage.

Short-Term Effects of Total Body Irradiation

Different tissues of the body respond differently to radiation, which is due to varying degrees of radiosensitivity. When an adult subject is irradiated over the entire body, various syndromes are manifested, depending on the dose applied. The effects of radiation are characterized by the survival time of the species and various stages of acute syndromes following whole-body irradiation.

Survival time varies with mammal species depending on the individual radiosensitivity. The radiosensitivity of a given species is commonly characterized by the lethal dose, $LD_{50/30}$, which is the dose that kills 50% of the species in 30 days. The $LD_{50/30}$ for humans is 250 to 300 rad (250–300 cGy); for dogs, 300 rad (300 cGy); and for mice, 900 rad (900 cGy).

Acute radiation syndromes appear in four stages: prodromal stage, latent stage, manifest illness stage, and recovery or death stage. Each stage is dose-dependent and can last for a few minutes to weeks. A minimum of 200 to 300 rad (200–300 cGy) is required for all four stages to be seen and can cause death.

In the prodromal stage, major symptoms are nausea, vomiting, and diarrhea, and occur in the early phase, lasting for only a short period of time, depending on the dose. A dose of 50 rad can induce nausea and vomiting. In the latent stage, biologic damage slowly builds up without any manifestation of syndromes, again lasting for hours to weeks, depending on the dose.

During the manifest illness stage, radiation syndromes appear as a result of the damage to the organ system involved, and the subject becomes ill. In the last stage, the subject either recovers or dies.

There are three categories of death due to radiation exposure depending on the dose: hemopoietic or bone marrow death, gastrointestinal (GI) death, or central nervous system (CNS) death.

Hemopoietic Death

Hemopoietic or bone marrow death occurs at a total body dose of 250–500 rad (250–500 cGy) in about 10–21 days following irradiation. At this dose, the precursors for RBCs and white blood cells are greatly affected, so much so that they lose the ability to reproduce. Loss of blood cell counts can be noticed at a dose as low as 10 to 50 rad (10 to 50 cGy). The number of lymphocytes is greatly depressed, whereby the immune system of the body is suppressed. Thus, the body loses defense against bacterial and viral infection and becomes susceptible to them. Immunosuppression by radiation occurs at doses as low as 100 rad (100 cGy) and 90% to 95% of immunosuppression can take place in humans at doses of 200–400 rad (200–400 cGy).

At this dose level, the platelet count is drastically reduced, and therefore bleeding gradually progresses through various orifices, which is due to lack of ability of the blood to coagulate. Fever, bleeding, and infection result, followed by ultimate death in 10–21 days.

Gastrointestinal Death

Gastrointestinal (GI) death occurs in 3–5 days at a whole-body dose of 500–10,000 rad (500–10,000 cGy). Following a large radiation dose, GI tract cells are destroyed and are not replaced, whereby an ulcer develops. Loss of proteins and nutrients through ulcers, in combination with the bacterial infection and excessive bleeding, results in GI death.

Central Nervous System Death

Central nervous system (CNS) death takes place in hours at a whole-body dose of more than 10,000 rad (10,000 cGy). Because the nerve cells are most radioresistant, such a large dose is required. At this dose level, malfunction of the neuron sodium pump, giving rise to motor incoordination, ataxia, intermittent stupor, and cardiovascular failure occur, followed ultimately by death.

Long-Term Effects of Radiation

The long-term or late effects of radiation cause various syndromes long after the radiation exposure. These may appear after acute radiation syndromes subside following exposure to a single large dose or after exposure to many smaller doses over a long period. The late effects may be somatic or genetic, depending on the respective cells involved. Somatic effects are seen in the form of carcinogenesis, life-shortening, cataractogenesis, and embryologic damage. On the other hand, genetic effects result in abnormalities in the offspring.

Somatic Effects

Carcinogenesis

It is well-known that long exposure to ionizing radiations can induce cancer in many species. Based on the data from the survivors of Hiroshima and Nagasaki, it has been estimated that there is a nominal risk of all cancers of 10^{-4}/rem (10^{-2}/Sv); that is, there is one in 10,000 chance of developing fatal cancer in humans resulting from total body exposure of 1 rem (0.01 Sv) at low doses (ICRP, 1977). Recently, a higher value of risk of 3 to 4×10^{-4} per rem (3 to 4×10^{-2} per Sv) has been reported (BEIRV, 1990). The incidence of leukemia and breast cancer due to low-level ionizing radiations is estimated to be 10–25 and 50–200 per rem per 10^6 persons, respectively.

The risk versus dose relationship has been controversial, particularly about the minimum level of radiation dose that induces cancer (Murphy, 1991). Some experts propose that the dose–response relationship is linear, indicating that there is no threshold dose below which there is no risk of cancer and that a very minimal dose can cause cancer (Fig. 12.6). A *threshold dose* is a hypothetical dose below which no radiation damage takes place in a population. Another group of experts considers that the risk–dose relationship is linear quadratic, meaning that it is linear at low doses and quadratic at

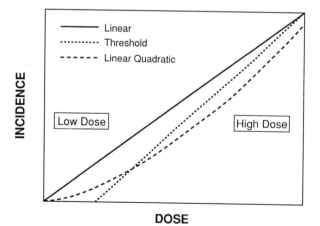

Fig. 12.6. Three general shapes of the dose–response curves permit prediction of different incidences of low-dose radiation effects when the curves are fitted to high-dose data. (Adapted from Murphy PH. Acceptable risk as a basis for regulation. *Radiographics* 1991;11:889–897.)

higher doses. Still other experts suggest that there is no risk of carcinogenesis up to a threshold dose, after which the curve becomes linear or quadratic.

The latent period of malignancies varies with the type of malignancy and the absorbed dose. Leukemia has an average latent period of about 5–10 years, whereas solid tumors in the head, neck, pharynx and thyroid have a latent period of more than 20–30 years.

Malignancies have been observed in individuals who are exposed to radiation from medical treatment, radiation-related occupation (e.g., industrial exposure), and acts of war. Infants and children are more radiosensitive than adults.

In the early 1900s, radium-dial painters used to lick the brush bristle soaked with radium-containing paint to make a fine point for painting dials of the clocks and watches. During the procedure, they ingested radium, which, as a chemical analogue of calcium, localized in bone, causing bone tumors. In some cases, the quantity of radium ingested was large, and acute effects including death were observed.

Before the 1930s, the enlarged thymus gland of an infant with acute respiratory distress syndromes was commonly treated with therapeutic doses of x-rays to reduce the enlargement. During irradiation, however, the thyroid glands also received a massive radiation dose. A statistically significant number of these infants developed thyroid cancer later in life (about 10 years later).

Radiologists who employed x-rays in their profession were found to have higher incidence of leukemia than other medical professionals. Dentists had higher incidence of finger lesions due to exposure to dental x-rays. These incidences occurred mostly before the 1950s, largely because of lack of knowledge of radiation effects. Now, through better radiation protection practice, these incidences have been curtailed drastically.

Increased incidences of leukemia, lung cancer, breast cancer, and thyroid cancer have been observed in the survivors of the atomic bomb attack on Hiroshima and Nagasaki.

Uranium mine workers inhale a considerable amount of radioactive dust containing radon gas. The decay products of radon settle in the lungs, and radiations from them can cause lung cancer.

Nonspecific Life-Shortening

Studies have shown that exposure to ionizing radiations results in the shortening of the life span of mice (Rotblat and Lindop, 1961). That is, the group irradiated with ionizing radiations dies sooner than the control group. The irradiated group looks much older than the control group, and radiation effects are similar to those of normal aging, e.g., an increase in connective tissue and a decrease in parenchymal cells. During the period of 1945–1955, American Radiologists were found to have shorter life-span than others medical professionals. But the issue of life-shortening by radiation is controversial, because in some cases it has been found that life span is rather lengthened by irradiation.

Cataractogenesis

The lens of the eye is very sensitive to radiation and develops cataract on irradiation with ionizing radiations. The incidence of radiation-induced cataract depends on the dose given. A dose of 10–30 rad (10–30 cGy) can produce cataract in mice, whereas a dose of 100 rad (100 cGy) is needed for humans.

Embryologic Damage

The developing mammalian embryo is extremely sensitive to ionizing radiations, because many cells are differentiating at this stage. The degree of damage however, depends on the developmental stage of the embryo. The entire fetal development is divided into three general stages: (1) *preimplantation*, a period of about 8–10 days between fertilization of the egg and its attachment to the uterine wall; (2) *major organogenesis*, a period of about 3–6 weeks, when major organs are developed; (3) *the fetal stage*, the remainder of the pregnancy period, when the organs of the fetus grow further to enable the mammal to survive after birth.

The embryo in the preimplantation stage is most sensitive to ionizing radiations and mostly encounters prenatal death as a result of radiation exposure. In some species, a dose of as low as 5–15 rad (5–15 cGy) is sufficient to cause deleterious effects on the embryo. Almost all embryos that survive the radiation exposure grow normally in utero and afterward, with the exception of a few who develop abnormalities later.

During the period of major organogenesis, embryos exposed to ionizing radiations develop abnormalities mostly related to the CNS and bone. These abnormalities are too severe for the fetus to survive and ultimately result in neonatal death. At an exposure of 200 rad (200 cGy) to mouse embryos during this period, almost 70% of the embryos later experienced neonatal death. Growth retardation also is noted at doses above 100 rad (100 cGy). Often it is suggested that a therapeutic abortion should be considered if an embryo receives ~ 10 rad (10 cGy) during the first 6 weeks after conception.

During the fetal period, however, comparable doses do not cause any abnormality or neonatal death. This is because fetal cells are more radioresistant than embryonic cells. Relatively higher doses are needed to cause death in this period. A few cases of growth retardation have been noted. In utero irradiation with a dose of 1–2 rad (1–2 cGy) may increase the risk of childhood leukemia by a factor of about 2. Structural and functional abnormalities of the CNS are the most frequent results of embryonic or fetal irradiation at a dose of as low as 10–20 rad (10–20 cGy).

Because of these radiation effects, practitioners should exercise caution in determining the woman's status of pregnancy before radiologic procedures.

If the woman is pregnant, then the risk versus benefit from the procedure to the patient should be weighed by the practitioner with due consideration to the stage of pregnancy. In normal patient scheduling, not knowing the age of pregnancy, one should implement the so-called 10-day rule, according to which the female patients of childbearing age are scheduled for all nuclear medicine procedures and for radiologic procedures involving the abdomen and pelvis only during the first 10 days after the onset of the menstrual cycle. This is considered the *safe period* because fertilization does not occur in this period.

Genetic Effects

As mentioned, ionizing radiations can induce mutations in the DNA structure, which ultimately are expressed in gene mutations. Through germ cells, these mutations propagate to future generations. Genetic effects are not expressed in the individual whose germ cells have been affected by radiation, but are expressed in future generations. Genetic effects appear as stillbirths, premature deaths after birth, sterility, and offspring with chromosome disorders.

Spontaneous Mutations

In normal cells, genes occasionally undergo natural mutations even without radiation exposure. Such mutations are called *spontaneous mutations*, and their frequency is about 10^{-5} per gene per generation. This means that the chance of spontaneous mutation is 1 in 100,000. This frequency is increased by various mutagens such as chemicals and radiation.

In a given generation, radiation does not produce any new mutations and simply increases the frequency of spontaneous mutations. Estimates are that approximately 10^{-7} mutations are produced per gene per rem. The dose–response relationship is linear, indicating that no dose is safe and that any dose, however small, causes some genetic mutations. Furthermore, genetic damage is a function of the dose rate and the LET of ionizing radiations. High-LET radiations and higher dose rates cause more mutations. Genetic mutations may appear in future generations long after exposure has occurred. A dose of more than 300–500 rad (300–500 cGy) can cause permanent sterility in humans.

Doubling Dose

A quantity called the *doubling dose* is a measure of the increase in genetic mutations by radiation. It is the amount of radiation dose that doubles the spontaneous mutations in a species. In humans, it is considered to be of the order of 50–250 rad (50–250 cGy), but it depends on the dose rate, the gender, and the type of species.

Genetically Significant Dose

The genetically significant dose (GSD) is the dose that, if received by everyone of the entire population, would cause the same genetic damage as the gonadal dose now being received by a limited number of individuals of the population through medical procedures, natural radiations, TV viewing, flying at high altitudes, and so forth. The GSD is an index of the expected genetic damage on the whole population and is calculated as an average value from the gonadal doses received from all exposures by the exposed personnel with proper weighting with respect to the chances of their having offspring. The weighting factor is needed because older people have lesser probability of having offspring than younger people.

The contributions of various sources of radiation to GSD are given in Table 12.3. The GSD values from natural radiation sources are considered to be equal to the gonadal dose, because natural radiations expose the entire population of all ages uniformly. Diagnostic x-rays contribute most to the GSD of all medical procedures. It is, therefore, essential that strict protective measures are taken to avoid unnecessary gonadal exposure. Gonadal shields, appropriate collimation of the x-ray beams, and limited and prudent application of repeat procedures definitely lead to a considerable reduction in GSD from medical procedures.

Genetic effects of radiation can be greatly reduced if a time interval is allowed between radiation exposure and conception. It is, therefore, recommended for both men and women that conception should be deferred for 6

Table 12.3. Annual genetically significant dose (GSD) in the U.S. population, about 1980–1982.

Source	Contributions to GSD in mrem (mSv)
Natural sources	
Radon	10 (0.1)
Other	90 (0.9)
Medical	
Diagnostic x-rays	20–30 (0.2–0.3)
Nuclear medicine	2 (0.02)
Consumer products	5 (0.05)
Occupational	~0.6 (0.006)
Nuclear fuel cycle	<0.05 (0.0005)
Miscellaneous environmental sources	<0.1 (0.001)
Total	~132 (1.32)

Adapted with permission from NCRP Report No. 93. *Ionizing Radiation Exposure of the Population of the United States.* Bethesda, Md: NCRP; 1987: Table 8.2.

months to 1 year after a significant radiation exposure such as a radiation accident or radiation therapy involving high gonadal exposure.

Questions

1. (a) What are the mechanisms of radiation damage?
   (b) Is the direct action or indirect action more contributing to radiation damage? Explain why.
   (c) Which are the free radicals that are most damaging to cells?
   (d) Does the presence of oxygen increase or decrease radiation damage?
2. (a) Why are erythroblasts more radiosensitive than red blood cells?
   (b) Which phase of the cell cycle is most radiosensitive?
   (c) Which molecule of the cell structure is most radiosensitive?
   (d) What are the different factors affecting radiation damage?
3. (a) Define D_0, D_q, and n as illustrated in the cell survival curve.
   (b) D_q is smaller for high-linear energy transfer (LET) radiations than low-LET radiations. True or False?
   (c) D_0 is smaller for high-LET radiations than for low-LET radiations. True or False?
   (d) How does LET affect the extrapolation number n?
   (e) What is the value of n for mammalian cells?
4. (a) The cell survival curve is steeper at high radiation doses than at low radiation doses. Explain why and its implication.
   (b) Does the shape of the cell survival curve vary with high-LET radiations and at very high dose rates?
5. (a) Choose the dose in rad that has been suggested as a practical threshold for radiation-induced abortion: (i) 2, (ii) 5, (iii) 10, (iv) 20, (v) 50.
   (b) What is the meaning of the 10-day rule for diagnostic x-ray examinations? Should this rule be applied to nuclear medicine studies?
   (c) How many days after conception can prenatal death occur as a result of in utero irradiation?
   (d) Which one of the following organs is most affected—malformed— by prenatal radiation exposure? (i) heart, (ii) stomach, (iii) head, (iv) gonads, (v) upper extremities.
   (e) What is the period of pregnancy when the incidence of abnormalities and malformations in human neonates is expected to be the highest?
   (f) What are the effects of radiation on the fetus?
   (g) The incidence of childhood leukemia after in utero irradiation with a few rad of diagnostic x-rays increases by a factor of: (i) 1.5–2, (ii) 2.0–3.0, (iii) 3.0–6.0.
6. (a) What are the dose ranges and approximate time limits for hemopoietic, gastrointestinal, and central nervous system death?
   (b) What are the prodromal syndromes, and when do they appear?

(c) What is the dose at which almost total immunosuppression occurs in humans?

7. (a) Define the oxygen enhancement ratio (OER).
   (b) Why is the oxygen effect absent for high-LET radiations?
   (c) What are radiosensitizers? Name some of them.
   (d) What is the maximum OER for γ- and x-rays?
   (e) Why is misonidazole a great radiosensitizer for hypoxic cells?
   (f) What is the specific composition of a radioprotector, and how does it function?

8. (a) What is the doubling dose, and what is its value for humans?
   (b) What is the dose at which permanent sterility can be induced in humans?
   (c) Define the genetically significant dose (GSD).
   (d) What is the GSD for humans?
   (e) Which one of all medical radiations contributes most to the GSD?
   (f) What are the factors that influence the GSD?

9. (a) The mean latent period for radiation-induced leukemia is about (i) 5–10 years, (ii) 10–20 years, (iii) 20–30 years.
   (b) The mean latent period for radiation-induced solid tumors is about (i) 5–10 years, (ii) 10–20 years, (iii) 20–30 years.
   (c) Cataract can be induced in humans with (i) 10–30 rad (10–30 cGy), (ii) 50–70 rad (50–70 cGy), (iii) 100–110 rad (100–110 cGy).
   (d) What is the risk of cancer in the general population from small doses of low-LET radiation exposure?

References and Suggested Readings

BEIR V Committee. *The Effects on Populations of Exposure to Low Levels of Ionizing Radiations.* Washington DC: National Academy of Sciences/National Research Council; 1990.

Hall EJ. *Radiobiology for the Radiologist.* 3rd ed. Philadelphia: JB Lippincott; 1988.

ICRP Report No. 26. *Recommendations of the International Commission on Radiological Protection.* New York: Pergamon; 1977.

Mettler FA, Moseley RD. *Medical Effects of Ionizing Radiations.* Orlando, Fla: Grune & Stratton; 1985.

Murphy PH. Acceptable risk as a basis for regulation. *Radiographics* 1991;11:889–897.

NCRP Report No. 93. *Ionizing Radiation Exposure of the Population of the United States.* Bethesda, Md: NCRP; 1987.

Pizzarello DJ, Witcofski RL. *Medical Radiation Biology.* 2nd ed. Philadelphia: Lea & Febiger; 1982.

Rotblat J, Lindop P. Long-term effects of a single whole body exposure of mice to ionizing radiation, II. Causes of death. *Proc R Soc Lond (Biol)* 1961;154:350–368.

Travis EL. *Primer of Medical Radiobiology.* Chicago: Year Book Medical Publishers, Inc; 1975.

CHAPTER 13

Internal Radiation Dosimetry

In Chapter 12, the effects of radiation on humans were discussed. These effects depend on various factors, such as dose, dose rate, time of exposure, and so on. In this chapter, we will describe the method of calculating absorbed doses in various organs from radionuclides ingested internally, either purposely (e.g., medical procedures) or accidentally. The calculation of radiation dose from internally absorbed radionuclides is detailed next.

Dose Calculation

The radiation absorbed dose depends on several factors: (1) the amount of radioactivity administered, (2) the physical and biological half-lives of the radioactivity, (3) the fractional abundance of the radiation in question from the radionuclides, (4) the biodistribution of radioactivity in the body, and (5) the fraction of energy released from the source organ that is absorbed in the target volume, which is related to the shape, composition, and location of the target. The physical characteristics of a radionuclide are well established. Information concerning the biodistribution of ingested radioactivity can be obtained from various experimental studies in humans and animals. Factors 4 and 5 vary from one individual to another, and therefore, they are approximated for a "standard" or "average" 70-kg man.

Radiopharmaceuticals administered to patients are distributed in different regions of the body. A region of interest for which the absorbed dose is to be calculated is considered the "target", whereas all other regions contributing to the radiation dose to the target are considered "sources". The source and the target become the same when the radiation dose due to the radioactivity in the target is calculated.

Radiation Dose Rate

Suppose a source volume r contains A μCi of a radiopharmaceutical emitting several radiations. If the ith radiation has energy E_i and a fractional abundance N_i per disintegration, then the energy absorbed per hour by a target of mass m and volume v from the ith radiation emitted by the source volume r (dose rate) is given by

$$R_i(\text{rad/h}) = A/m(\mu\text{Ci/g})N_iE_i \ (\text{MeV/disintegration})$$

$$\times \ [3.7 \times 10^4 \ \text{disintegrations}/(\text{s} \cdot \mu\text{Ci})]$$

$$\times \ (1.6 \times 10^{-6} \ \text{erg/MeV})$$

$$\times \ (0.01 \ \text{g} \cdot \text{rad/erg})$$

$$\times \ (3600 \ \text{s/hr})$$

$$= 2.13(A/m)N_iE_i$$

If the target and the source are not the same, then a factor must be introduced to account for the partial absorption, if any, of the radiation energy. Thus,

$$R_i(\text{rad/hr}) = 2.13(A/m)N_iE_i\Phi_i(v \leftarrow r) \tag{13.1}$$

Here, $\Phi_i(v \leftarrow r)$ is called the *absorbed fraction* and is defined as the ratio of the energy absorbed by the target volume v from the ith radiation to the energy emitted by the ith radiation from the source volume r. This is a critical factor that is difficult to evaluate because the absorbed fraction Φ_i depends on the energy of the radiation, the shape and size of the source volume, and the shape, composition, and distance of the target volume. However, in the case of β particles, conversion electrons, α particles, and x- and γ-rays of energies less than 11 keV, all of the energy emitted by a radionuclide is absorbed in the volume r larger than 1 cm. Then, Φ_i becomes 0, unless v and r are the same, in which case $\Phi_i = 1$. For x- and γ-rays with energies greater than 11 keV, the value of Φ_i decreases with increasing energy and varies between 0 and 1, depending on the energy. The values of Φ_i are calculated by statistical methods on the basis of fundamental mechanisms of interaction of radiation with matter and are available in many standard textbooks on radiation dosimetry, particularly the medical internal radiation dose (MIRD) pamphlets published by the Society of Nuclear Medicine.

The quantity $2.13N_iE_i$ is a constant for the ith radiation and is often denoted by Δ_i. Thus,

$$\Delta_i = 2.13N_iE_i \tag{13.2}$$

The quantity Δ_i is called the *equilibrium dose constant* for the ith radiation and has the unit $\text{g} \cdot \text{rad}/(\mu\text{Ci} \cdot \text{hr})$ based on the units chosen in Eq. (13.1). Note that β particles are emitted with a distribution of energy and therefore the average energy, $\bar{E}_\beta$, of β particles is used in the calculation of Δ_i. Thus, Eq. (13.1) becomes

$$R_i(\text{rad/hr}) = (A/m)\Delta_i\Phi_i(v \leftarrow r) \tag{13.1}$$

The activity A will change as a result of the physical decay and biological elimination of the radiopharmaceutical, and therefore the dose rate will also change. Assuming an effective exponential change in A with time, Eq. (13.3) can be written

$$R_i(\text{rad/hr}) = (A_0/m)\Delta_i e^{-\lambda_e t}\Phi_i(v \leftarrow r) \tag{13.4}$$

Here, λ_e is the effective decay constant of the radiopharmaceutical, and t is the time over which the original activity A_0 has decayed.

Cumulative Radiation Dose

The cumulative radiation dose D_i to the target due to the ith radiation of the radionuclide during the period $t = 0$ to t can be obtained by integrating Eq. (13.4). Thus,

$$D_i(\text{rad}) = \frac{A_0}{m}\Delta_i\Phi_i(v \leftarrow r) \int_0^t e^{-\lambda_e t} dt$$

$$= \frac{A_0}{m}\Delta_i\Phi_i(v \leftarrow r)\frac{1}{\lambda_e}(1 - e^{-\lambda_e t})$$

$$= 1.44\frac{A_0}{m}\Delta_i T_e(1 - e^{-\lambda_e t})\Phi_i(v \leftarrow r) \tag{13.5}$$

Here, T_e is the effective half-life of the radiopharmaceutical in hours (discussed in Chapter 3). If $t = \infty$, that is, if the radiopharmaceutical is completely eliminated by physical and biological decay, then the exponential term $e^{-\lambda_e t}$ approaches zero and the absorbed dose in Eq. (13.5) may be written as

$$D_i(\text{rad}) = 1.44(A_0/m)\Delta_i T_e\Phi_i(v \leftarrow r) \tag{13.6}$$

If the radionuclide has n radiations with energies $E_1, E_2, \ldots E_n$ and fractional abundances $N_1, N_2, \ldots N_n$ per disintegration, then the total dose D can be obtained by summing Eq. (13.6) over all radiations. Thus,

$$D(\text{rad}) = 1.44\frac{A_0}{m}T_e\sum_{i=1}^n \Delta_i\Phi_i(v \leftarrow r) \tag{13.7}$$

This summation can also be applied to Eq. (13.4) for the dose rate R_i. The total dose to the target from other sources can be calculated by summing Eq. (13.7) over all sources.

In the MIRD pamphlets, the values of Δ_i have been compiled on the basis of various nuclear characteristics of the radionuclide in question. The Φ_i values have been calculated on the basis of different sizes and compositions of the targets receiving the radiation dose and the radiation characteristics of the radionuclide. In MIRD pamphlet No. 11 (Snyder et al., 1975) Eq. (13.7) has been substituted by

$$D(\text{rad}) = \tilde{A} \cdot S \tag{13.8}$$

where

$$\tilde{A} = 1.44 \times A_0 \times T_e \qquad (13.9)$$

$$S = \Sigma\Delta_i\Phi_i/m \qquad (13.10)$$

The quantity $\tilde{A}$ is called the *cumulated activity* and has the unit of $\mu Ci \cdot hr$. The quantity S is called the *mean absorbed dose* per cumulated activity and has the unit of $rad/\mu Ci \cdot hr$. The values of S are tabulated in MIRD pamphlet No. 11. The MIRD dose estimate reports are available for many commonly used radiopharmaceuticals and are published periodically by the Society of Nuclear Medicine.

Problem 13.1
Calculate the absorbed dose to the lungs from the administration of 4 mCi (148 MBq) of ^{99m}Tc-labeled macroaggregated albumin (MAA) particles, assuming that 99% of the particles are trapped in the lungs. The value of S for the lungs is 5.2×10^{-5} rad/$\mu Ci \cdot hr$. Assume that the ^{99m}Tc activity is uniformly distributed in the lungs and that 40% of the activity is cleared from the lungs with a biological half-life of 3 hr and 55% with a half-life of 7 hr.

Answer
The half-life of ^{99m}Tc = 6 hr
The effective half-life of two biological clearances are

$$T_{e1} = \frac{3 \times 6}{3 + 6} = 2 \text{ hr}$$

$$T_{e2} = \frac{7 \times 6}{7 + 6} = 3.2 \text{ hr}$$

Using Eq. (13.9)

$$\tilde{A} = 1.44 \times 4000 \times 0.99 \times (0.45 \times 2 + 0.55 \times 3.2)$$

$$= 15{,}200 \; \mu Ci \cdot hr \; (0.562 \text{ GBq} \cdot hr)$$

Using Eq. (13.10)

$$D = \tilde{A} \cdot S$$

$$= 15{,}200 \times 5.2 \times 10^{-5}$$

$$= 0.79 \text{ rad}$$

$$= 790 \text{ mrad} \; (7.9 \text{ mGy})$$

Radiation Dose in SI Units

The radiation dose in System Internationale (SI) units due to the administration of a radiopharmaceutical can be calculated by assuming a source volume r containing A MBq of the radiopharmaceutical that emits several radiations. If the ith radiation has energy E_i and a fractional abundance N_i per disintegration, then the energy absorbed per hour by a target of mass m and volume v from the ith radiation emitted by the source volume r (dose rate) is given by

$$R_i(\text{Gy/hr}) = A/m(\text{MBq/g})N_iE_i(\text{MeV/disintegration})$$

$$\times \, 10^6 \, \text{disintegrations}/(\text{s} \cdot \text{MBq})$$

$$\times \, (1.6 \times 10^{-6} \, \text{erg/MeV})$$

$$\times \, (1 \times 10^{-4} \, \text{g} \cdot \text{Gy/erg})$$

$$\times \, (3600 \, \text{s/hr})$$

$$= 0.576 \, (A/m)N_iE_i$$

When the target and the source are not the same, the absorbed fraction $\Phi(v \leftarrow r)$ must be taken into account. Thus,

$$R_i(\text{Gy/hr}) = 0.576(A/m)N_iE_i\Phi(v \leftarrow r) \qquad (13.11)$$

The quantity $0.576N_iE_i$ is a constant and can be denoted by Δ_i, as in Eq. (13.2). Thus,

$$\Delta_i = 0.576N_iE_i \qquad (13.12)$$

With this value of Δ_i, Eqs. (13.3)–(13.8) are equally applicable to radiation doses in SI units. It should be understood that the equations in SI units contain a constant $\Delta_i = 0.576N_iE_i$ and activities expressed in MBq, whereas the equations in conventional units contain the equilibrium dose constant $\Delta_i = 2.13N_iE_i$ and activities expressed in microcuries.

Table 13.1 presents radiation absorbed doses from various radiopharmaceuticals to different organs in adults.

Pediatric Dosages

The metabolism, biodistribution, and excretion of drugs differ in children from those in adults, and therefore radiopharmaceutical dosages for children must be adjusted. Several methods and formulas have been reported on pediatric dosage calculations based on body weight, body surface area, combination of weight and area, and simple ratios of adult dosages. The calculation based on body surface area is more accurate for pediatric dosages. The body surface area of an average adult is 1.73 m² and proportional to the 0.7 power of the body weight. Based on the information, the Paediatric Task Group European Association Nuclear Medicine Members published the fractions of the adult dosages needed for children, which are shown in Table 13.2. For most nuclear studies, however, there is a minimum dosage required for a meaningful scan, which is primarily established in each institution based on experience.

Table 13.1. Radiation-absorbed doses in adults for various radiopharmaceuticals.

Radiopharmaceutical	Organ	Dose	
		rad/mCi	mGy/GBq
^{99m}Tc-pertechnetate*	Whole body	0.011	3.0
	Thyroid	0.130	35.1
	Upper large intestine	0.120	32.4
	Lower large intestine	0.110	30.0
	Stomach	0.051	13.8
	Ovaries	0.030	8.1
	Testes	0.009	2.4
^{99m}Tc-sulfur colloid[†]	Whole body	0.016	4.3
	Liver	0.380	102.7
	Spleen	0.210	56.8
	Marrow	0.028	7.6
^{99m}Tc-DTPA[‡]	Whole body	0.016	4.3
	Bladder	0.450	121.6
	Kidneys	0.040	10.8
	Gonads	0.015	4.0
^{99m}Tc-MAA[§]	Whole body	0.015	4.0
	Lungs	0.280	75.6
	Kidneys	0.160	43.2
	Liver	0.080	21.6
	Ovaries	0.018	4.9
	Testes	0.015	4.0
^{99m}Tc-MDP[‖]	Whole body	0.007	1.9
	Bone	0.038	10.3
	Bladder (wall)	0.440	118.9
	Kidneys	0.031	8.4
	Marrow	0.025	6.8
	Ovaries	0.017	4.6
	Testes	0.012	3.2
^{99m}Tc-DISIDA[¶]	Whole body	0.016	4.3
	Liver	0.039	10.5
	Lower large intestine	0.390	105.4
	Upper large intestine	0.270	73.0
	Gallbladder	0.120	32.4
^{99m}Tc-sestamibi[#]	Whole body	0.017	4.6
	Gallbladder	0.067	18.1
	Upper large intestine	0.100	27.0
	Lower large intestine	0.180	48.6
	Heart (wall)	0.017	4.6
	Kidneys	0.067	18.1
	Ovaries	0.053	14.3
	Bladder (wall)	0.140	37.8
^{99m}Tc-MAG3[#]	Whole body	0.007	1.9
	Bladder wall	0.480	129.7
	Kidneys	0.014	3.8
	Ovaries	0.026	7.0
^{99m}Tc-HMPAO[#]	Whole body	0.013	3.5
	Brain	0.026	7.0
	Thyroid	0.100	27.0
	Kidneys	0.130	35.1
	Gallbladder	0.190	51.4
	Lachrymal gland	0.258	69.7

Table 13.1. (*continued*)

Radiopharmaceutical	Organ	Dose	
		rad/mCi	mGy/GBq
^{131}I-iodide$^{\#}$	Whole body	0.45	121.6
	Thyroid	1300.00	3.5×10^5
	Liver	0.48	130.00
^{123}I-iodide**	Whole body	0.03	8.1
	Thyroid	13.00	3513.5
	Testes	0.02	5.4
^{82}RbCl$^{\#}$	Whole body	0.002	0.5
	Kidneys	0.032	8.6
	Heart (wall)	0.007	1.9
^{201}Tl-thallous chloride$^{\#}$	Whole body	0.21	56.8
	Heart	0.40	135.1
	Kidneys	1.20	324.3
	Liver	0.55	148.6
	Testes	0.50	135.1
^{18}F-FDG††	Whole body	0.04	10.8
	Heart	0.16	43.2
	Bladder	0.44	118.9
	Spleen	0.16	43.2
^{67}Ga-gallium citrate‡‡	Whole body	0.26	70.2
	Liver	0.46	124.3
	Marrow	0.58	156.7
	Spleen	0.53	143.2
	Upper large intestine	0.56	151.4
	Lower large intestine	0.90	243.2
	Gonads	0.26	70.2

* Data from MIRD Dose Estimate Report No. 8. *J Nucl Med.* 1976; 17:74.

† Data from MIRD Dose Estimate Report No. 3. *J Nucl Med.* 1975; 16(1): 108A–B.

‡ Data from Hauser H, Atkins HL, Nelson KG, et al. Technetium-99m-DTPA: a new radiopharmaceutical for brain and kidney imaging. *Radiology.* 1970; 94:679.

§ Data from Robbins PJ, Feller PA, Nishiyama H. Evaluation and dosimetry of a ^{99m}Tc-Sn-MAA lung imaging agent in humans. *Health Physics.* 1976; 30: 173.

$^{\parallel}$ Data from Subramanian G, McAfee JG, Blair RJ, et al. Technetium-99m-methylene diphosphonate—a superior agent for skeletal imaging: comparison with other agents. *J Nucl Med.* 1975; 16:744.

¶ Data from Snyder WS, Ford MR, Warner GG, et al. "S" Absorbed dose per unit cumulated activity for selected radionuclides and organs. *MIRD Pamphlet No. 11*, New York; Society of Nuclear Medicine, 1975.

$^{\#}$ Data from package inserts of the respective product.

** Data from Kereiakes JG, Feller PA, Ascoli FA, et al. Pediatric radiopharmaceutical dosimetry. In: *Proceedings of Radiopharmaceutical Dosimetry Symposium*, Oak Ridge, Tenn; April 1976.

†† Data from Jones SC, Alavi A, Christman D, et al. The radiation dosimetry of 2-[F-18]Fluoro-2-Deoxy-D-glucose in man. *J Nucl Med.* 1982; 23:613.

‡‡ Data from MIRD Report No. 2. *J Nucl Med.* 1973; 14:755.

Table 13.2. Fraction of adult-administered dosages for pediatric administration.

Weight in kg (lb)	Fraction	Weight in kg (lb)	Fraction
3 (6.6)	0.10	28 (61.6)	0.58
4 (8.8)	0.14	30 (66.0)	0.62
8 (17.6)	0.23	32 (70.4)	0.65
10 (22.0)	0.27	34 (74.8)	0.68
12 (26.4)	0.32	36 (79.2)	0.71
14 (30.8)	0.36	38 (83.6)	0.73
16 (35.2)	0.40	40 (88.0)	0.76
18 (39.6)	0.44	42 (92.4)	0.78
20 (44.0)	0.46	44 (96.8)	0.80
22 (48.4)	0.50	46 (101.2)	0.83
24 (52.8)	0.53	48 (105.6)	0.85
26 (57.2)	0.56	50 (110.0)	0.88

Adapted from Paediatric Task Group European Association Nuclear Medicine Members. A radiopharmaceuticals schedule for imaging paediatrics. *Eur J Nucl Med.* 1990;17:127.

Questions

1. Calculate the absorbed dose to the thyroid gland of a hyperthyroid patient from a dosage of 30 mCi of ^{131}I, assuming 60% uptake, a biologic half-life of 4 days for thyroid clearance of ^{131}I, and S equal to 2.2×10^{-2} rad/μCi·hr.
2. Calculate the dose in rem and sieverts to a tumor that received 35 rad (0.35 Gy) from neutron therapy (quality factor = 10 for neutrons).
3. What is the usual whole-body radiation dose from most nuclear medicine studies using ^{99m}Tc-labeled radiopharmaceuticals?
4. Identify as true or false whether the following affect the absorption fraction of a γ-emitting radionuclide:
   (a) γ-Ray energy
   (b) Shape of the target organ
   (c) Composition of the target organ
   (d) Amount of the radioactivity present in the source
   (e) Shape of the source organ
5. Does the mean absorbed dose per cumulated activity, S, depend on:
   (a) Absorbed fraction
   (b) Target mass
   (c) Photon energy
   (d) Photon abundance
6. What is the important parameter that is considered in adjusting the activity to be administered to children compared to adults for a nuclear medicine test?
7. Calculate the cumulated activity $\tilde{A}$ in a 55-g source organ containing 3 mCi(111 MBq) of ^{99m}Tc($t_{1/2}$ = 6 hr) with a biological $t_{1/2}$ = 14 hr.

8. A target organ has a mass of 35 g and contains 1 mCi (37 MBq) of a radionuclide emitting a β^- particle with $\Delta_1 = 0.3$ g·rad/μCi·hr and $\Phi_1 = 1.0$, and a γ-radiation with $\Delta_2 = 0.2$ g·rad/μCi·hr and $\Phi_2 = 0.35$. Calculate the mean absorbed dose per cumulated activity.

9. An external beam deposits 360 ergs of energy in 3 g of tissue. What is the radiation dose in rad?

References and Suggested Readings

Fourth International Pharmaceutical Dosimetry Symposium. CONF-85113. November 1985; Oak Ridge, Tenn.

International Commission on Radiological Protection. Radiation dose to patients from radiopharmaceuticals. *ICRP 53*. New York: Pergamon Press; 1988.

Kereiakes JG, Rosenstein M. *Handbook of Radiation Doses in Nuclear medicine and Diagnostic X-ray*. Boca Raton, Fla: CRC Press; 1980.

Snyder WS, Ford MS, Warner GG, et al. "S" absorbed dose per unit cumulated activity for selected radionuclides and organs. *MIRD Pamphlet No. 11*. New York: Society of Nuclear Medicine; 1975.

Snyder WS, Ford MR, Warner GG. Specific absorbed fractions for radiation sources uniformly distributed in various organs of a heterogeneous phantom. *MIRD Pamphlet No. 12*. New York: Society of Nuclear Medicine; 1977.

CHAPTER 14

Radiation Protection and Regulations

Radiation Protection

Since the discovery of radioactivity, there has been a tremendous increase in the use of radionuclides for various purposes. At the same time, radiation hazards have also increased considerably. To minimize radiation hazards, international and national organizations have been established to set guidelines for the safe handling of radioactive materials. The International Committee on Radiation Protection (ICRP) and the National Council on Radiation Protection and Measurement (NCRP) are two such organizations. They set guidelines for all radiation workers to follow in handling radiation. The Nuclear Regulatory Commission (NRC) adopts many of these recommendations into regulations for implementing radiation protection programs in the United States. At present, the *Federal Register, Code of Federal Regulations,* 10CFR20 contains major radiation protection guidelines applicable in the United States. The 10CFR20 has been revised, and the revision will be effective as of January 1, 1994. This revision incorporates advances in science and new concepts of radiation protection and includes all principal recommendations of the ICRP.

Definition of Terms

Because the revised 10CFR20 will be implemented soon, this discussion refers to the revised 10CFR20 and omits the old 10CFR20. There are major changes in regulations concerning radiation protection in the revised 10CFR20. Because it is beyond the scope of this book to include the entire 10CFR20, only the relevant highlights are included.

Several terms related to absorbed dose as defined in the revised 10CFR20 are given here.

Committed dose equivalent ($H_{T,50}$) is the dose equivalent to organs or tissues of reference (T) that will be received from an intake of radioactive material by an individual during the 50-year period following the intake.

Table 14.1. Values of weighting factor.

Organ or Tissue	W_T
Gonads	0.25
Breast	0.15
Red bone marrow	0.12
Lung	0.12
Thyroid	0.03
Bone surfaces	0.03
Remainder	0.30
Whole body	1.00

From *Federal Register. Code of Federal Regulations*. 10CFR20 (revised). Washington, DC: U.S. Government Printing Office; 1991.

Deep-dose equivalent (H_d), which applies to the external whole-body exposure, is the dose equivalent at a tissue depth of 1 cm (1000 mg/cm^2).

Shallow-dose equivalent (H_s), which applies to the external exposure of the skin or an extremity, is the dose equivalent at a tissue depth of 0.007 cm (7 mg/cm^2) averaged over an area of 1 cm^2.

Weighting factor (W_T) for an organ or tissue is the proportion of the risk of stochastic effects resulting from irradiation of that organ or tissue to the total risk of stochastic effects when the whole body is irradiated uniformly. The values of W_T from the 10CFR20 are shown in Table 14.1.

Effective dose equivalent (H_E) is the sum of the products of the weighting factors applicable to each of the body organs or tissues that are irradiated and the committed dose equivalent to the corresponding organ or tissue ($H_E = \sum W_T \cdot H_{T,50}$).

Annual limit on intake (ALI) is the derived limit on the amount of radioactive material allowed to be taken into the body of an adult worker by inhalation or ingestion in a year. These values are given in 10CFR20 (Table 1, Appendix B).

Total effective dose equivalent (TEDE) is the sum of the deep-dose equivalent (for external exposure) and the committed effective dose equivalent (for internal exposure).

Radiation area is an area in which an individual could receive from a radiation source a dose equivalent in excess of 5 mrem (0.05 mSv) in 1 hr at 30 cm from the source.

High-radiation area is an area where an individual could receive from a radiation source a dose equivalent in excess of 100 mrem (1 mSv) in 1 hr at 30 cm from the source.

Very high-radiation area is an area where an individual could receive from radiation sources an absorbed dose in excess of 500 rad (5 Gy) in 1 hr at 1 m from the source.

Sources of Radiation Exposure

The population at large receives radiation exposure from various sources such as natural radioactivity, medical procedures, cosmetic products, and occupational sources. The estimates of various average annual exposures are tabulated in Table 14.2.

The major contribution of the exposure comes from natural sources, particularly from radon from building materials, amounting to 200 mrem (2 mSv)/year. Excluding radon exposure, the average exposure from natural background consisting of cosmic radiations, terrestrial radiations, and so on amounts to about 100 mrem (1 mSv)/year. This exposure varies with the altitude of places above sea level. For example, the annual cosmic ray exposure in cities such as Denver is about 50 mrem (0.5 mSv) compared to 26 mrem (0.26 mSv) at sea level. Air travel at a height of 39,000 ft (12 km) gives 0.5 mrem/hr (5 μSv/hr), resulting in an annual dose of 1 mrem (10 μSv) to the population.

Terrestrial radiation exposure arises from radionuclides such as ^{40}K and

Table 14.2. Annual effective dose equivalent in the U.S. population from different sources circa 1980 to 1982.

Sources	Average annual effective dose equivalent in mrem (mSv)
Natural sources	
Radon	200 (2.0)
Cosmic rays	27 (0.27)
Terrestrial	28 (0.28)
Ingested radionuclides	39 (0.39)
Medical procedures	
Diagnostic x-rays	39 (0.39)
Nuclear medicine	14 (0.14)
Radiation therapy	<1 (0.01)
Consumer products	5–13 (0.05–0.13)
Occupational	0.9 (0.009)
Nuclear fuel cycle	0.05 (0.0005)
Miscellaneous	0.06 (0.0006)
Total	~360 (3.6)

Adapted with permission from NCRP Report No. 93. *Ionizing Radiation Exposure of the Population of the United States.* Bethesda, Md.: NCRP; 1987: Table 8.2.

from decay products of thorium and uranium in soil. It varies from about 16 mrem (160 μSv)/year in the Atlantic sea to 63 mrem (630 μSv)/year in the Rockies with an average of 28 mrem (280 μSv)/year.

Radionuclides ingested through food, water, or inhalation include ^{40}K and decay products of thorium and uranium, particularly ^{210}Po, and contribute about 39 mrem (390 μSv) annually.

Medical procedures contribute the highest exposure of all man-made radiation sources. The most exposure comes from diagnostic radiographic procedures with about 39 mrem (390 μSv) annually compared to 14 mrem (140 μSv) for nuclear medicine procedures. Exposure from radiation therapy is relatively small.

Consumer products such as tobacco, water, building materials, agricultural products, and television receivers contribute to radiation exposure through consumption. Exposure from smoking has been estimated to be 1.3 mrem (13 μSv)/year, which is not included in Table 14.2, because it is difficult to calculate the collective effective dose equivalent for the entire population. The total exposure from consumer products varies between 5 and 13 mrem (50 and 130 μSv)/year.

Occupational exposure is received by the workers in reactor plants, coal mines, and other industries using radionuclides. This value is about 0.9 mrem (9 μSv)/year, which is quite small, because a great deal of precaution is taken to reduce exposure in the workplace.

Nuclear power plants around the country release small amounts of radionuclides to the environment, which cause radiation exposure to the population. This value is of the order of 0.05 mrem (0.5 μSv)/year.

Radiation Dose Limits

The annual limit of the occupational dose to an individual adult is the more limiting of (1) TEDE of 5 rem (50 mSv) or (2) the sum of the deep-dose equivalent and the committed dose equivalent to any individual organ or tissue other than the lens of the eye being equal to 50 rem (0.5 Sv). There is no lifetime cumulative dose limit in the revised 10CFR20.

The annual limit on the occupational dose to the lens of the eye is 15 rem (150 mSv).

The annual limit of the occupational dose to the skin and other extremities is the shallow-dose equivalent of 50 rem (0.5 Sv).

The annual occupational dose limits for minors are 10% of the annual dose limits for adults. The dose limit to the fetus or embryo during the entire pregnancy from occupational exposure of a declared pregnant woman is 0.5 rem (5 mSv).

The total effective dose equivalent to individual members of the public is 0.1 rem (1 mSv)/year. This limit, however, can be increased to 0.5 rem (5 mSv)/year, provided the need for such a higher limit is demonstrated.

The dose in an unrestricted area from an external source is 2 mrem (20 μSv)/hr and 50 mrem (0.5 mSv)/year.

Principles of Radiation Protection

Of the various types of radiation, the α particle is most damaging because of its charge and great mass, followed in order by the β particle and the γ-ray. Heavier particles have shorter ranges and therefore deposit more energy per unit path length in the absorber, causing more damage. These are called *nonpenetrating radiations*. On the other hand, γ-rays and x-rays have no charge or mass and therefore have a longer range in matter and are called *penetrating radiations*. These radiations cause relatively less damage in tissue. Knowledge of the type and energy of radiations is essential in understanding the principles of radiation protection.

The cardinal principles of radiation protection from external sources are based on four factors: time, distance, shielding, and activity.

Time

The total radiation exposure to an individual is directly proportional to the time of exposure to the radiation source. The longer the exposure, the higher the radiation dose. Therefore, it is wise to spend no more time than necessary near radiation sources.

Distance

The intensity of a radiation source, and hence the radiation exposure, varies inversely as the square of the distance from the source to the point of exposure. It is recommended that an individual should keep as far away as possible from the radiation source. Procedures and radiation areas should be designed so that individuals conducting the procedures or staying in or near the radiation areas receive only minimum exposure.

The radiation exposure from γ-ray and x-ray emitting radionuclides can be estimated from the *exposure rate constant*, Γ, which is defined as the exposure from γ-rays and x-rays in R/hr from 1 mCi (37 MBq) of a radionuclide at a distance of 1 cm. Each γ- and x-ray emitter has a specific value of Γ, which has the unit of R $\cdot$ cm^2/mCi $\cdot$ hr at 1 cm or, in System Internationale (SI) units, μGy $\cdot$ m^2/GBq $\cdot$ hr at 1 m. The Γ values are derived from the number of γ-ray and x-ray emissions from the radionuclide, their energies, and their absorption coefficients in air. Because γ-rays or x-rays below some 10 or 20 keV are absorbed by the container and thus do not contribute significantly to radiation exposure, often, γ-rays and x-rays above these energies only are included in the calculation of Γ. In these instances, they are denoted by Γ_{10} or Γ_{20}. The values of Γ_{20} for different radionuclides are given in Table 14.3.

The exposure rate X from an n-mCi radionuclide source at a distance d cm

Table 14.3. Exposure rate constants of commonly used radionuclides.

Radionuclides	Γ_{20} (R·cm^2/mCi·hr at 1 cm)	Γ_{20} (μGy·m^2/GBq·hr at 1 m)*
^{137}Cs	3.26	88.11
^{99m}Tc	0.59	15.95
^{201}Tl	0.45	12.16
^{99}Mo	1.46	39.46
^{67}Ga	0.76	20.54
^{123}I	1.55	41.89
^{111}In	2.05	55.41
^{125}I	1.37	37.03
^{57}Co	0.56	15.16
^{131}I	2.17	58.65
^{18}F†	5.70	154.05

*R·cm^2/mCi·hr is equal to 27.027 μGy·m^2/GBq·hr.
†Personal communication with Dr. M. Stabin, Oak Ridge Associated Universities, Inc., Oak Ridge, Tennessee.
Adapted from Goodwin PN: Radiation safety for patients and personnel. In: Freeman LM, ed. *Freeman and Johnson's Clinical Radionuclide Imaging.* 3rd ed. Philadelphia: WB Saunders Co; 1984:320.

is given by

$$X = \frac{n\Gamma}{d^2} \tag{14.1}$$

where Γ is the exposure rate constant of the radionuclide.

Problem 14.1
Calculate the radiation exposure at 25 cm from a vial containing 30 mCi (1.11 GBq) of ^{201}Tl.

Answer
The exposure rate constant Γ_{20} of ^{201}Tl is 0.45 R·cm^2/mCi·hr at 1 cm from Table 14.3. Therefore, using Eq. (14.1), at 25 cm

$$X = \frac{30 \times 0.45}{25^2} = 21.6 \text{ mR/hr}$$

Because Γ_{20} of ^{201}Tl in SI units is 12.16 μGy·m^2/GBq·hr at 1 m, X for 1.11 GBq of ^{201}Tl at 25 cm is

$$X = \frac{1.11 \times 12.16}{(0.25)^2}$$

$$= 215.96 \ \mu\text{Gy/hr}$$

Shielding

Various high atomic number (Z) materials that absorb radiations can be used to provide radiation protection. Because the ranges of α and β particles are short in matter, the containers themselves act as shields for these radiations. γ-radiations, however, are highly penetrating. Therefore, highly absorbing material should be used for shielding of γ-emitting sources, although for economic reasons, lead is most commonly used for this purpose.

Obviously, shielding is an important means of protection from radiation. Radionuclides should be stored in a shielded area. The radiopharmaceutical dosages for patients should be carried in shielded syringes. Radionuclides emitting β particles should be stored in containers of low-Z material such as aluminum and plastic because in high-Z material, such as lead, they produce highly penetrating bremsstrahlung radiations. For example, ^{32}P should be stored in plastic containers instead of lead containers.

Activity

It should be obvious that the radiation exposure increases with the intensity of the radioactive source. The greater the source strength, the more the radiation exposure. Therefore, one should not work unnecessarily with large quantities of radioactivity.

Personnel Monitoring

According to the revised 10CFR20, personnel monitoring is required under the following conditions:

1. Occupational workers including minors and pregnant women likely to receive in 1 year a dose in excess of 10% of the annual limit of exposure from the external radiation source
2. Individuals entering high or very high radiation areas

Monitoring for occupational intake of radioactive material is also required if the annual intake by an individual is likely to exceed 10% of the ALIs in 10CFR20, Table 1, Appendix B, and if the minors and pregnant women are likely to receive a committed effective dose equivalent in excess of 0.05 rem (0.5 mSv) in 1 year.

Three devices are used to measure the exposure of ionizing radiations received by an individual: the pocket dosimeter, the film badge, and the thermoluminescent dosimeter. The pocket dosimeter has been described in Chapter 7.

The film badge is most popular and cost-effective for personnel monitoring and gives reasonably accurate readings of exposures from β-, γ- and x-radiations. The film badge consists of a radiation-sensitive film held in a

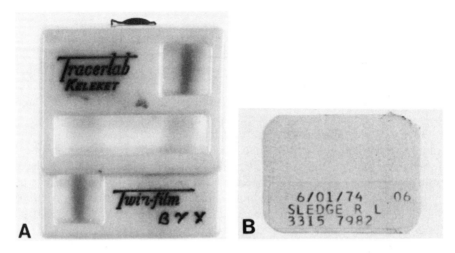

Fig. 14.1. (A) Film badge holder. (B) Film badge.

plastic holder (Fig. 14.1). Filters of different materials (aluminum, copper, and gold) are attached to the holder in front of the film to differentiate exposure from radiation of different types and energies. After exposure the optical density of the developed film is measured by a densitometer and compared with that of a calibrated film exposed to known radiation. Film badges are usually changed monthly for each radiation worker in most institutions. Film badges provide an integral dose and a permanent record. The main disadvantage of the film badge is the long waiting period before the exposed personnel know about their exposure. The film badge also tends to develop fog resulting from heat and humidity, particularly when in storage for a long time, and this may obscure the actual exposure reading.

In many institutions the film badges of all workers are sent to a commercial firm that develops and reads the density of the films and sends the report of exposure to the institution. According to the revised 10CFR20, the history of previous radiation exposure is not required because there is no limit for lifetime cumulative dose.

A thermoluminescent dosimeter (TLD) consists of inorganic crystals (chips) such as lithium fluoride (LiF) and manganese-activated calcium fluoride (CaF_2 : Mn) held in holders like the film badges and plastic rings. When these crystals are exposed to radiation, electrons from the valence band are excited and trapped by the impurities in the forbidden band. If the radiation-exposed crystal is heated to 300°C to 400°C, the trapped electrons are raised to the conduction band; they then fall back into the valence band, emitting light. The amount of light emitted is proportional to the amount of radiation absorbed in the TLD. The amount of light is measured and read as the amount of radiation exposure by a TLD reader, a unit that heats the crystal and reads the exposure as well. The TLD gives an accurate exposure reading and can be reused after proper heating (annealing).

It should be noted that exposure resulting from medical procedures and background radiations are not included in occupational dose limits. Therefore, radiation workers should wear film badges or dosimeters only at work. These devices should be taken off during any medical procedures involving radiation such as radiographic procedures and dental examinations and also when leaving after the day's work.

Dos and Don'ts in Radiation Protection Practice

Do wear laboratory coats and gloves when working with radioactive materials.

Do work in a ventilated fume hood while working with volatile material.

Do cover the trays and workbench with absorbent paper.

Do store and transport radioactive material in lead containers.

Do wear a film badge while working in the radiation laboratory.

Do identify all radionuclides and dates of assay on the containers.

Do survey work areas for contamination as frequently as possible.

Do clean up spills promptly and survey the area after cleaning.

Do not eat, drink, or smoke in the radiation laboratory.

Do not pipette any radioactive material by mouth.

Do monitor hands and feet after the day's work.

Do notify the radiation safety officer (RSO) in the case of any major spill or other emergencies related to radiation.

Radiation Regulations

Nuclear Regulatory Commission

Currently, in the United States the NRC regulates all reactor-produced by-product materials with regard to their use and disposal and the radiation safety of all personnel using them as well as the public. The NRC does not regulate naturally occurring and accelerator-produced radionuclides that are regulated by the individual states. At the time of this writing, 29 states, in agreement with the NRC, are authorized to regulate the reactor-produced by-product materials in addition to the naturally occurring and accelerator-produced materials. These states are called Agreement States.

License

Authorization for the use of radioactive materials is granted by issuance of a license by the NRC or the Agreement State. There are two types of licenses:

1. *General license:* The general license is granted to physicians, veterinarians, clinical laboratories, and hospitals for specific purposes (10CFR31.11)

without the filing of an application with the NRC. The total amount of activity to be possessed should not exceed 200 μCi (7.4 MBq) of ^{131}I, ^{125}I, ^{75}Se, ^{59}Fe, ^{14}C, or ^{3}H.

2. *Specific licenses:* The specific licenses are of two types: specific license of limited scope and specific license of broad scope (10CFR33). The specific license of limited scope is issued to physicians in private practice or to a medical institution for specific use of limited quantities of by-product materials in humans. A radiation safety officer is required under this license.

The specific license of broad scope for medical use is issued only to larger medical institutions for multiple types and large quantities of by-product materials for unspecified uses. The radiation safety committee and the radiation safety officer are required to supervise the use and disposal of radioactive materials within the institution. Individual users with training and experience are approved by the radiation safety committee for specific protocols using radioactive material.

Radiation Safety Committee

The NRC requires in the medical institutional license that a radiation safety committee (RSC) be established, consisting of at least a user for each type of use permitted by the license, the RSO, a representative of the nursing staff, and a representative of the management. Additional members may be included at the institution's option. The committee is charged with responsibility for evaluating all proposals for research, diagnostic, and therapeutic uses of radionuclides; the committee reviews and recommends the radiation safety procedures according to the ALARA program (see later) in the institution and must meet at least quarterly.

Radiation Safety Officer

An RSO is appointed to implement the radiation safety program as implied in the license of the institution. The RSO's activities include investigations of overexposures, accidents, spills, losses, thefts, misadministrations and unauthorized receipts, uses and transfers, and instituting corrective actions. The RSO sets, with the recommendation of the RSC, policies and procedures for purchase, receipt, storage, transfer, and disposal of all radioactive materials within the institution. The RSO conducts periodic checks of survey instruments, dose calibrators, surveys of radiation areas, and records all activities related to radiation in the institution.

ALARA Program

The established dose limits are the upper limits for radiation exposure to individuals. The NRC has instituted the ALARA (as low as reasonably achievable) concept to reduce radiation exposure to individuals to a minimum. The ALARA concept calls for a reasonable effort to maintain individual and collective radiation exposure as low as possible. Under this concept, techniques, equipment, and procedures are all critically evaluated. According to NRC Regulatory Guide 10.8, Appendix G, under the ALARA concept, when the exposure to a radiation worker exceeds 10% of the occupational exposure limit in a quarter (Action Level I), an investigation is made by the RSO, and the report is reviewed by the RSC. When the exposure exceeds 30% of the occupational exposure limit (Action Level II), corrective actions are taken or the licensee must justify a higher dose level for ALARA in that particular situation.

Medical Uses of Radioactive Materials

There are six categories of medical uses of radioactive materials according to 10CFR35. They are (1) radiopharmaceuticals for uptake, dilution, and excretion (10CFR35.100); (2) radiopharmaceuticals for imaging and localization including the use of generators and kits (10CFR35.200); (3) radiopharmaceuticals for therapy (10CFR35.300); (4) sealed sources for brachytherapy (10CFR35.400); (5) sealed sources for diagnosis such as sources of ^{125}I and ^{153}Gd for bone mineral analysis (10CFR35.500); and (6) sealed sources for teletherapy such as sources of ^{60}Co and ^{137}Cs in teletherapy units (10CFR35.600). The guidelines for the medical use of all radioactive materials are given in 10CFR35, but only radiopharmaceuticals under categories 1, 2, and 3 are relevant in nuclear medicine.

Survey for Contamination and Exposure Rate

The NRC requires a daily survey of all areas where radiopharmaceuticals are prepared or administered, a weekly survey of areas where radiopharmaceuticals or radioactive wastes are stored, and a weekly wipe testing of areas where radiopharmaceuticals are prepared, administered, or stored (10CFR35.70). Area surveys for ambient exposure rate are conducted with a Geiger–Müller (G–M) survey meter. Wipe tests for external contamination with radioactivity are performed by swabbing several 100-cm^2 spot areas using absorbent paper and counting them in a NaI(Tl) well counter. Wipe tests for removable contamination must be sensitive enough to detect as low as 2000 disintegrations per minute. The permissible contamination for ^{131}I usually is 0.001 μCi (37Bq)/100 cm^2 and that for ^{99m}Tc is 0.01 μCi (370Bq)/100 cm^2 (NRC Regulatory Guide 10.8, Appendix N).

Dose Calibrators and Survey Meters

According to the NRC regulations, a dose calibrator is required to radio-assay radiopharmaceuticals (10CFR35.50) and a survey meter is required for area survey (10CFR35.120 and 10CFR35.220). The calibration of the dose calibrators is described in Chapter 7. The survey meters are calibrated annually. The portable radiation detection survey meter should be capable of detecting dose rates over the range of 0.1 mrem/hr (1 μSv/hr) to 100 mrem/hr (1 mSv/hr), and the portable radiation measurement survey meter should cover a range of 1 mrem/hr (10 μSv/hr) to 1000 mrem/hr (10 mSv/hr).

Caution Signs and Labels

The NRC requires that specific signs, symbols, and labels be used to warn people of possible danger from the presence of radiation. These signs use magenta, purple, or black color on yellow background; some typical signs are shown in Figure 14.2.

Caution: Radiation Area. This sign must be posted in radiation areas.

Caution: High Radiation Area or *Danger: High-Radiation Area.* This sign must be posted in high-radiation areas.

Caution: Radioactive Material or *Danger: Radioactive material.* This sign is posted in areas or rooms in which 10 times the quantity of any licensed material specified in Appendix C of 10CFR20 are used or stored. All containers with quantities of licensed materials exceeding those specified in Appendix C of 10CFR20 should be labeled with this sign. These labels must be removed or defaced before disposal of the container in the unrestricted areas.

Caution signs are not required in rooms storing the sealed sources, provided the radiation exposure at 1 foot (30 cm) from the surface of the source

Fig. 14.2. Various radiation caution signs and labels.

reads less than 5 mrem (50 μSv)/hr. Caution signs are not needed in rooms where radioactive materials are handled for less than 8 hr, during which time the materials are constantly attended.

Bioassay

NRC regulatory guide 8.20 gives the details of bioassay requirements for ^{131}I and ^{125}I radionuclides. Bioassays are required when the level of radioiodine activity handled (volatile or dispersible) exceeds the following values:

Open bench: 1 mCi (37 MBq)
Fume hood: 10 mCi (370 MBq)
Glove box: 100 mCi (3.7 GBq)

When the radioiodinated material is nonvolatile, the limits of activity are higher by a factor of 10. Stricter limits may be imposed in the license by the NRC.

For iodine radionuclides, bioassay is performed by the thyroid uptake test within 72 hr and at 14 days after handling the radioactivity. Sometimes urine analysis may also be required soon after the exposure. Bioassays may be required for other radionuclides, depending on the amount and type of radionuclides.

Receiving and Monitoring of Radioactive Packages

Individual users or institutions are authorized to possess and use radioactive materials on issuance of a radioactive material license by the NRC or the Agreement State Agency. The suppliers require documentation of licensing of the user as to the types and limits of quantities of radioactive material before shipping.

Monitoring of packages is required if the packages are labeled as containing radioactive material or if the packages are damaged or leaking. A radioactive shipment must be monitored as soon as possible after receipt but no later than 3 hr after delivery if the delivery takes place during normal hours, or not later than 3 hr from the beginning of the next working day if it is received after working hours. Two types of monitoring are performed: survey for external exposure and wipe test for contamination on the surface of the package resulting from leakage of liquid. The survey reading of external exposure should not exceed 200 mrem/hr (2 mSv/hr) on the surface of the container or 10 mrem/hr (100 μSv/hr) at 1 m from the surface of the container. The wipe test should show less than 22,000 dpm or 370 Bq/100 cm^2. If the readings exceed these limits, the NRC and the final delivering carrier must be notified by telephone and telegram, mailgram, or facsimile. Advice should be sought from these authorities as to whether the shipment should be returned.

After all surveys are completed, the data must be entered into a receipt book. The information logged in includes the date of the receipt, the manufacturer, the lot number, name and quantity of the product, date and time of calibration, and survey data along with the initials of the individual processing the receipt.

Radioactive Waste Disposal

Radioactive waste generated in nuclear medicine or pharmacy (e.g., syringes, needles, vials containing residual activities and contaminated papers, tissues, and liners) are disposed of by the following methods according to the guidelines set forth in the revised 10CFR20, which takes effect in January 1994:

1. Decay in storage
2. Release into a sewerage system
3. Transfer to authorized recipient (commercial land disposal facilities)
4. Other disposal methods approved by the NRC (e.g., incineration of solid waste and atmospheric release of radioactive gases)

The following is a brief description of different methods of radioactive waste disposal, but one should consult the revised 10CFR20 for details.

Decay in Storage

Although the revised 10CFR20 does not spell out the conditions of the decay-in-storage method, 10CFR35.92 describes this method in detail. Radionuclides with half-lives less than 65 days usually are disposed of by this method. These radionuclides are allowed to decay in storage for a minimum of 10 half-lives and then surveyed. If the radioactivity of the waste cannot be distinguished from background, it can be disposed of in the normal trash after removal or defacing of all radiation labels. This method is most appropriate for short-lived radionuclides such as ^{99m}Tc, ^{123}I, ^{201}Tl, ^{111}In, and ^{67}Ga. Radioactivities should be stored separately according to half-lives for convenience of timely disposal of each radionuclide.

Release into Sewerage System

The NRC permits radioactive waste disposal into the sewerage system provided the radioactive material is soluble or dispersible in water and the quantity disposed monthly does not exceed the maximum permissible limits set in 10CFR20. Disposal depends on the total volume and flow rate of water used but is limited to 1 Ci (37 GBq) of ^{14}C, 5 Ci (185 GBq) of ^{3}H, and 1 Ci (37 GBq) of all other radionuclides annually. Excreta from humans undergoing medical diagnosis or treatment with radioactive material are exempted from these limitations. However, items contaminated with radioactive excreta (e.g., linen, diapers, etc., contaminated with urine or feces) are not exempted from these limitations. To adopt this method of radioactive disposal, one must determine the total volume and the flow of sewer water in the institution and the number of users of a specific radionuclide so that for each individual user, a limit can be set for sewer disposal of the radionuclide in question.

Transfer to Authorized Recipient

This method of transfer to an authorized recipient is adopted for long-lived radionuclides and usually involves transfer of radioactive wastes to authorized commercial firms that bury or incinerate at approved sites or facilities. This method is facing a serious problem because the existing facilities may close down soon, and many states are not developing a solution.

Although the columns of the ^{99}Mo–^{99m}Tc generators may be decayed to background for disposal to normal trash, a convenient method of disposing of this generator is to return them to the vendors, who let them decay and later dispose of them. Normally, the generator is picked up by the authorized carrier when a new one is delivered.

Other Disposal Methods

A licensee may adopt methods of radioactive waste disposal different from those mentioned here, provided regulatory agency approval is obtained. Impact of such disposal methods on environment, nearby facilities, and population is heavily weighed before approval. Incineration of solid radioactive waste and carcasses of research animals containing radioactive materials is allowed by this method. Radioactive gases such as ^{133}Xe and ^{127}Xe are also released by this method, as long as their maximum permissible concentration at the effluent side of the exhaust to the atmosphere does not exceed the NRC limits. Radioactive waste containing 0.05 μCi (1.85 kBq) or less of ^{3}H or ^{14}C/g of medium used for liquid scintillation counting or animal tissue may be disposed of in the regular nonradioactive trash.

Records must be maintained as to the date of storage and the amount and kind of activity stored in a waste disposal log book. The stored packages must be labeled with pertinent information. The date of disposal and the amount of disposed activity must also be recorded in the log book, along with the initials of the individual disposing of the waste.

Radioactive Spill

Accidental spillage of radioactivity can cause unnecessary radiation exposure to personnel and must be treated cautiously and expeditiously. Appropriate procedures must be established for handling radioactive spills. There are two types of spills: major spill and minor spill. No definitive distinction exists between a minor and a major spill. A major spill usually occurs when the spilled activity cannot be contained in a normal way and can cause undue exposure to personnel. In the case of a major spill, the RSO should be notified immediately. In either case, the access to the area should be restricted.

Areas, personnel, and equipment must be decontaminated, keeping in mind the principle of containment of radioactivity. Survey and wipe tests must be performed after decontamination. The RSO will investigate the accident and recommend corrective action if a major spill occurs.

Record Keeping

Records must be maintained for the receipt, storage, and disposal of radioactive materials, as well as for various activities performed in the radiation laboratories. According to the NRC regulations, these records must contain specific information and must be kept for a certain period of time specified by the NRC.

Transportation of Radioactive Materials

The transportation of radioactive materials is governed by the U.S. Department of Transportation (DOT), which establishes the guidelines for packaging, types of packaging material, limits of radioactivity in a package, and exposure limits. Title 49 of the *Code of Federal Regulations* (49CFR) and 10CFR71 contain all these regulations related to packaging and transportation of radioactive materials.

There are two types of packaging:

Type A: This type of packaging is used primarily for most radiopharmaceuticals. Such packaging is sufficient to prevent loss of radioactive material with proper shielding to maintain the prescribed exposure during normal transportation. The limits of radioactivities of various radionuclides under this category are specified in 49CFR and 10CFR71.

Type B: When the radioactivity exceeds the limits specified in Type A, Type B packaging must be used. Such packaging is considerably more accident resistant and is required for very large quantities of radioactive material.

The packages must pass certain tests such as the drop test, corner drop test, compression test, and 30-min water spray test.

The radioactive packages must be labeled properly before transportation. There are three types of labels (Fig. 14.3) according to the exposure reading in mR/hr at 1 m from the surface of the package (*transport index*). The criteria for the three labels are given in Table 14.4. The transport index (TI) must be indicated on the label and the sign "RADIOACTIVE" must be placed on the package. The maximum permissible TI value is 10, although it is limited to three for passenger-carrying aircrafts. For liquids, the label "THIS SIDE UP" must be placed on the package. Each package must be labeled on opposite sides with the appropriate warning label (one of the labels in Table 14.4). The label must identify the content and amount of radionuclide in curies or becquerels. The package must contain shipping documents inside, bearing

Fig. 14.3. The three types of U.S. Department of Transportation labels required for transportation of radioactive materials.

Table 14.4. Labeling categories for packages containing radioactive materials.

	Exposure (mR/hr)	
Type of label	At surface	At 1 m
White-I	<0.5	—
Yellow-II	>0.5 ≤ 50	<1
Yellow-III	> 50 ≤ 200	>1 ≤ 10

Note: No package shall exceed 200 mR/hr at the surface of the package or 10 mR/hr at 1 m. Transport index is the reading in mR/hr at 1 m from the package surface (10CFR71).

the identity, amount, and chemical form of the radioactive material and the TI.

Placards are necessary on the transport vehicles carrying yellow-III–labeled packages and must be put on four sides of the vehicle.

Questions

1. Define committed dose equivalent, deep-dose equivalent, total effective dose equivalent, radiation area, and high radiation area.
2. What are the annual dose limits for radiation workers for:
   (a) Whole body
   (b) Lens
   (c) Extremities
3. What is the dose limit in the unrestricted area and for the individual members of the public?

4. (a) Calculate the exposure rate at 10 inches from a 150-mCi (5.55-GBq)–
^{131}I source (Γ_{20} of ^{131}I = 2.17 R·cm^2/mCi·hr at 1 cm).
   (b) If the half-value layer (HVL) of lead for ^{131}I is 3 mm, how much lead is
needed to reduce the exposure to 10% of the calculated value at 10
inches?
5. Why is ^{32}P stored in plastic and not in lead containers?
6. What is the approximate amount of lead necessary to reduce the exposure
rate from a 200-mCi–^{99m}Tc source to less than 5 mR/hr at 20 cm from the
source? (Γ_{20} of ^{99m}Tc = 0.59 R·cm^2/mCi·hr at 1 cm = 15.95 μGy·m^2/
GBq·hr at 1 m; HVL of Pb for ^{99m}Tc = 0.3 mm).
7. If 1% of the primary beam exits through a patient, calculate the exposure
at the midline of the patient.
8. (a) Who are required to wear personnel monitoring devices?
   (b) Film badges can discriminate different types of radiation. True or
False?
   (c) Film badges can discriminate radiations of different energies. True or
False?
   (d) Why are filters used in film badges?
   (e) Filters convert radiation energies into visible light. True or False?
   (f) Filters protect the individual from radiation exposure. True or False?
   (g) Describe how thermoluminescent dosimeters work.
9. (a) What is the ALARA program?
   (b) What is an Agreement State?
   (c) How often should area surveys and wipe tests be performed in nuclear
medicine?
   (d) When does one take a bioassay?
   (e) What are the NRC requirements for survey of the packages on receipt?
   (f) Describe different methods of disposal of radioactive waste.
   (g) What are the general principles of handling radioactive spillage?
   (h) What is a transportation index (TI), and how is it used in the transportation of radioactive material?

References and Suggested Readings

Federal Register. Code of Federal Regulations. 10CFR20 (revised). Washington, DC:
U.S. Government Printing Office; 1991.

Federal Register. Code of Federal Regulations. 10CFR31. Washington, DC: U.S.
Government Printing Office; 1992.

Federal Register. Code of Federal Regulations. 10CFR33. Washington, DC: U.S.
Government Printing Office; 1992.

Federal Register. Code of Federal Regulations. 10CFR35. Washington, DC: U.S.
Government Printing Office; 1992.

Federal Register. Code of Federal Regulations. 10CFR71. Washington, DC: U.S.
Government Printing Office; 1992.

Federal Register. Code of Federal Regulations. 49CFR170 to 49CFR189. Washington, DC: U.S. Government Printing Office; 1989.

National Council on Radiation Protection and Measurements. *Basic Radiation Protection Criteria (NCRP 39).* Bethesda, Md: NCRP Publications; 1971.

National Council on Radiation Protection and Measurements. *Nuclear Medicine– Factors Influencing the Choice and Use of Radionuclides Diagnosis and Therapy (NCRP 70).* Bethesda, Md: NCRP Publications; 1982.

National Council on Radiation Protection and Measurements. *Ionizing Radiation Exposure of the Population of the United States (NCRP 93).* Bethesda, Md: NCRP Publications; 1987.

National Council on Radiation Protection and Measurements. *Radiation Protection and Allied Health Personnel (NCRP 105)* Bethesda, Md: NCRP Publications; 1989.

Quimby E. *Safe Handling of Radioactive Isotopes in Medical Practice.* New York: Macmillan Publishing Co; 1976.

Shapiro J. *Radiation Protection.* 2nd ed. Cambridge, Mass: Harvard University Press; 1981.

U.S. NRC Regulatory Guide 10.8. Washington, DC: U.S. Government Printing Office; 1987.

Appendix A

Units and Constants

Energy

1 electron volt (eV)	$= 1.602 \times 10^{-12}$ erg
1 kiloelectron volt (keV)	$= 1.602 \times 10^{-9}$ erg
1 million electron volts (MeV)	$= 1.602 \times 10^{-6}$ erg
1 joule (J)	$= 10^7$ ergs
1 watt (W)	$= 10^7$ ergs/s
	$= 1$ J/s
1 rad	$= 1 \times 10^{-2}$ J/kg
	$= 100$ ergs/g
1 gray (Gy)	$= 100$ rad
	$= 1$ J/kg
1 sievert (Sv)	$= 100$ rem
	$= 1$ J/kg
1 horsepower (HP)	$= 746$ W
1 calorie (cal)	$= 4.184$ J

Charge

1 electronic charge	$= 4.8 \times 10^{-10}$ electrostatic unit
	$= 1.6 \times 10^{-19}$ C
1 coulomb (C)	$= 6.28 \times 10^{18}$ charges
1 ampere (A)	$= 1$ C/s

Mass and Energy

1 atomic mass unit (amu)	$= 1.66 \times 10^{-24}$ g
	$= 1/12$ the atomic weight of ^{12}C
	$= 931$ MeV
1 electron rest mass	$= 0.511$ MeV
1 proton rest mass	$= 938.78$ MeV
1 neutron rest mass	$= 939.07$ MeV
1 pound	$= 453.6$ g

Length

1 micrometer, or micron (μm)	$= 10^{-6}$ m
	$= 10^4$ Å

1 nanometer (nm)	$= 10^{-9}$ m
1 angstrom (Å)	$= 10^{-8}$ cm
1 fermi (F)	$= 10^{-13}$ cm
1 inch	$= 2.54$ cm

Activity

1 curie (Ci)	$= 3.7 \times 10^{10}$ disintegrations per second (dps)
	$= 2.22 \times 10^{12}$ disintegrations per minute (dpm)
1 millicurie (mCi)	$= 3.7 \times 10^{7}$ dps
	$= 2.22 \times 10^{9}$ dpm
1 microcurie (μCi)	$= 3.7 \times 10^{4}$ dps
	$= 2.22 \times 10^{6}$ dpm
1 becquerel (Bq)	$= 1$ dps
	$= 2.703 \times 10^{-11}$ Ci
1 kilobecquerel (kBq)	$= 10^{3}$ dps
	$= 2.703 \times 10^{-8}$ Ci
1 megabecquerel (MBq)	$= 10^{6}$ dps
	$= 2.703 \times 10^{-5}$ Ci
1 gigabecquerel (GBq)	$= 10^{9}$ dps
	$= 2.703 \times 10^{-2}$ Ci
1 terabecquerel (TBq)	$= 10^{12}$ dps
	$= 27.03$ Ci

Constants

Avogadro's number	$= 6.02 \times 10^{23}$ atoms/g·atom
	$= 6.02 \times 10^{23}$ molecules/g·mole
Planck's constant (h)	$= 6.625 \times 10^{-27}$ erg·s/cycle
Velocity of light	$= 3 \times 10^{10}$ cm/sec
π	$= 3.1416$
e	$= 2.7183$

Appendix B

Terms Used in Text

Absorption. A process by which the total energy of a radiation is removed by an absorber through which it passes.

Accelerator. A machine to accelerate charged particles linearly or in circular paths by means of an electromagnetic field. The accelerated particles such as α particles, protons, deuterons, and heavy ions possess high energies and can cause nuclear reactions in target atoms by irradiation.

Accuracy. A term used to indicate how close a measurement of a quantity is to its true value.

Annihilation radiation. γ-Radiations of 511 keV energy emitted at 180° after a β^+ particle is annihilated by combining with an electron in matter.

Atomic mass unit (amu). By definition, one twelfth of the mass of $^{12}_{6}C$, equal to 1.66×10^{-24} g or 931 MeV.

Atomic number (Z). The number of protons in the nucleus of an atom.

Attenuation. A process by which the intensity of radiation is reduced by absorption and/or scattering during its passage through matter.

Attenuation coefficient. The fraction of γ-ray energy attenuated (absorbed plus scattered) per unit length of an absorber (linear attenuation coefficient, μ) or per gram of an absorber (mass attenuation coefficient, μ_m).

Auger electron. An electron ejected from an energy shell, instead of a characteristic x-ray, carrying the energy equal to that of the x-ray.

Average life (τ). *See* Mean life.

Avogadro's number. The number of molecules in 1 g·mole of any substance or the number of atoms in 1 g·atom of any element. It is equal to 6.02×10^{23}.

Binding energy. The energy to bind two entities together. In a nucleus, it is the energy needed to separate a nucleon completely from other nucleons in the nucleus. In a chemical bond, it is the energy necessary to separate two binding partners an infinite distance.

Biological half-life (T_b). The time by which one half of an administered dose of a substance is eliminated by biological processes such as urinary and fecal excretions.

Bremsstrahlung. γ-Ray photons produced by deceleration of charged particles near the nucleus of an absorber atom.

Carrier. A stable element that is added in detectable quantities to a radionuclide of the same element, usually to facilitate chemical processing of the radionuclide.

Carrier-free. A term used to indicate the absence of any stable atoms in a radionuclide sample.

Collimator. A device to confine a beam of radiation within a specific field of view. Collimators may be converging, pinhole, diverging, and parallel-hole types.

Collimator efficiency. The number of photons passing through the collimator for each unit of activity present in a source.

Collimator resolution. A component of spatial resolution of an imaging system contributed by the collimator. It is also called *geometric resolution.*

Committed dose equivalent ($H_{T,50}$). The dose equivalent to organs or tissues of reference (T) that will be received from an intake of radioactive material by an individual during the 50-year period following intake.

Compton scattering. In this process, a γ-ray transfers only a partial amount of energy to an outer orbital electron of an absorber, and the photon itself is deflected with less energy.

Conversion electron (e^-). *See* Internal conversion.

Critical organ. See Organ, critical.

Cross section (σ). The probability of occurrence of a nuclear reaction or the formation of a radionuclide in a nuclear reaction. It is expressed in a unit termed *barn*; 1 barn $= 10^{-24}$ cm^2.

Curie (Ci). A unit of activity. A curie is defined as 3.7×10^{10} disintegrations per second.

Dead time. The period of time that a counter remains insensitive to count the next after an event.

Decay constant (λ). The fraction of atoms of a radioactive element decaying per unit time. It is expressed as $\lambda = 0.693/t_{1/2}$, where $t_{1/2}$ is the half-life of the radionuclide.

Deep-dose equivalent (H_d). Dose equivalent at a tissue depth of 1 cm (1000 mg/cm^2) resulting from external whole-body exposure.

Dose. The energy of radiation absorbed by any matter. Also, a general term for the amount of a radiopharmaceutical administered in microcuries or millicuries.

Dosimeter. An instrument to measure the cumulative dose of radiation received during a period of radiation exposure.

Dosimetry. The calculation or measurement of radiation absorbed doses.

Effective half-life (T_e). Time required for an initial administered dose to be reduced to one half as a result of both physical decay and biological elimination of a radionuclide. It is given by $T_e = (T_p \times T_b)/(T_p + T_b)$, where T_e is the effective half-life, and T_p and T_b are the physical and biological half-lives, respectively.

Electron (e⁻). A negatively charged particle rotating around the atomic nucleus. It has a charge of 4.8×10^{-10} electrostatic units and a mass of 9.1×10^{-28} g, equivalent to 0.511 MeV, or equal to 1/1836 of the mass of a proton.

Electron capture (EC). A mode of decay of a proton-rich radionuclide in which an orbital electron is captured by the nucleus, accompanied by emission of a neutrino and characteristic x-rays.

Electron volt (eV). The kinetic energy gained by an electron when accelerated through a potential difference of 1 V.

Energy resolution. Capability of a detecting system to separate two γ-ray peaks of different energies. It is given by the full width at half maximum (FWHM) of a given photopeak.

Erg. The unit of energy or work done by a force of 1 dyne through a distance of 1 cm.

Fission (f). A nuclear process by which a nucleus divides into two nearly equal smaller nuclei, along with the emission of two to three neutrons.

Free radical. A highly reactive chemical species that has one or more unpaired electrons.

Generator, radionuclide. A device in which a short-lived daughter is separated chemically and periodically from a long-lived parent adsorbed on adsorbent material. For example, ^{99m}Tc is separated from ^{99}Mo from the Moly generator with saline.

Gray (Gy). The unit of absorbed radiation dose in SI units. One gray is equal to 100 rad.

Half-life ($t_{1/2}$). A unique characteristic of a radionuclide, defined by the time during which an initial activity of a radionuclide is reduced to one half. It is related to the decay constant λ by $t_{1/2} = 0.693/\lambda$.

Half-value layer (HVL). The thickness of an absorbing material required to reduce the intensity or exposure of a radiation beam to one half of the initial value when placed in the path of the beam.

Internal conversion. An alternative mode to γ-ray decay in which nuclear excitation energy is transferred to an orbital electron, which is then ejected from the orbit.

Ion. An atom or group of atoms with a positive charge (cation) or a negative charge (anion).

Intrinsic efficiency. The number of radiations detected divided by the number of radiations striking the detector.

Intrinsic resolution. A component of the spatial resolution of an imaging system that is contributed by the detector and associated electronics and depends on the photon energy, detector thickness, and the number of PM tubes.

Isobars. Nuclides having the same mass number, that is, the same total number of neutrons and protons. Examples are $^{57}_{26}$Fe and $^{57}_{27}$Co.

Isomeric transition (IT). Decay of the excited state of an isomer of a nuclide to a lower excited state or the ground state.

Isomers. Nuclides having the same atomic and mass numbers but differing in energy and spin of the nuclei. For example, ^{99}Tc and ^{99m}Tc are isomers.

Isotones. Nuclides have the same number of neutrons in the nucleus. For example, $^{131}_{53}$I and $^{132}_{54}$Xe are isotones.

Isotopes. Nuclides having the same atomic number, that is, the same number of protons in the nucleus. Examples are $^{14}_{6}$C and $^{12}_{6}$C.

$LD_{50/30}$. A quantity of a substance that, when administered or applied to a group of any living species, kills 50% of the group in 30 days.

Linear energy transfer (LET). Energy deposited by radiation per unit length of the matter through which the radiation passes. Its usual unit is keV/μm.

Mass defect. The difference between the mass of the nucleus and the combined masses of individual nucleons of the nucleus of a nuclide.

Mass number (A). The total number of protons and neutrons in a nucleus of a nuclide.

Mean life (τ). The average expected lifetime of a group of radionuclides before disintegration. It is related to the half-life and decay constant by $\tau = 1/\lambda = 1.44t_{1/2}$.

Metastable state (m). An excited state of a nuclide that decays to the ground state by the emission of γ-radiation with a measurable half-life.

Modulation transfer function. A quantitative value of the spatial resolution of an imaging system.

Neutrino (v). A particle of no charge and mass emitted with variable energy during β^{+}, and electron capture decays of radionuclides. An antineutrino ($\bar{v}$) is emitted in β^{-} decay.

No carrier added (NCA). A term used to characterize the state of a radioactive material to which no stable isotope of the compound has been added purposely.

Nucleon. A common term for neutrons or protons in the nucleus of a nuclide.

Organ, critical. The organ that is functionally essential for the body and receives the highest radiation dose after administration of radioactivity.

Organ, target. The organ intended to be imaged and expected to receive the greatest concentration of administered radioactivity.

Pair production. γ-rays with energy greater than 1.02 MeV interact with the nucleus of an absorber atom, and a positron and an electron are produced at the expense of the photon.

Photoelectric effect. A process in which a γ-ray, while passing through an absorber, transfers all its energy to an orbital electron, primarily the K-shell electron of an absorber, and the photoelectron is ejected from the shell.

Photofraction. The fraction of all detected γ-rays that contributes to the photopeak.

Physical half-life (T_{p}). *See* Half-life.

Precision. A term used to indicate the reproducibility of the measurement of a quantity when measurements are made repeatedly.

Quality factor (QF). A factor dependent on linear energy transfer that is multiplied by absorbed doses to calculate the dose equivalents in rem. It is used in radiation protection to take into account the relative radiation damage caused by different radiations. It is 1 for x-, γ-, and β-rays and 10 for neutrons and protons.

Rad. The unit of radiation-absorbed dose. One rad is equal to 100 ergs of radiation energy deposited per gram of any matter, or 10^{-2} J/kg of any matter.

Range. The straight line distance traversed by a charged particle in an absorber.

Relative biologic effectiveness (RBE). A factor used to calculate the dose equivalent in rem from rad. It is defined as the ratio of the amount of a standard radiation that causes certain biological damage to the amount of radiation in question that causes the same biological damage.

Roentgen. The quantity of x- or γ-radiations that produces one electrostatic unit of positive or negative charge in 1 cm^3 of air at 0°C and 760-mm Hg pressure (standard temperature and pressure, STP). It is equal to 2.58×10^{-4} C/kg air.

Roentgen equivalent man (rem). A dose equivalent defined by the absorbed dose (rad) times the relative biological effectiveness or quality factor of the radiation in question.

Scintillation scanning or imaging. Recording of the distribution of radioactivity in the body or a section of the body by the use of a NaI(Tl) detector to form an image.

Scintigraphy. A photographic recording of the distribution of radioactivity in an area of interest in the body by the use of a gamma camera.

Sensitivity. The number of counts per unit time detected by an imaging device for each unit of activity present in a source. It is expressed in cps/μCi.

Shallow-dose equivalent (H_s). Dose equivalent at a tissue depth of 0.007 cm (7 mg/cm^2) averaged over an area of 1 cm^2 from external exposure to the skin.

Sievert (Sv). The SI unit of dose equivalent and equal to 100 rem.

Spatial resolution. A measure of the ability of an imaging device to faithfully reproduce the image of an object. It is given by the modulation transfer function (MTF) and is determined by the Fourier transform of the line spread function.

Specific activity. The amount of radioactivity per unit mass of a radionuclide or labeled compound.

Specific ionization. The number of primary and secondary ion pairs produced by an incident radiation per unit path length in an absorber.

Thermal neutron. Neutrons of thermal energy 0.025 eV.

Appendix C

Answers to Questions

Chapter 2

3. 81.3%
7. 130 keV

Chapter 3

1. (a) 1.11×10^{15} atoms
   (b) 0.24 μg
2. (a) 4.75×10^{14} dpm
   (b) 216 Ci or 7.99×10^{12} Bq
3. 6.97 hr
4. (a) 429 mCi (15.9 GBq)
   (b) 120.7 mCi (4.46 GBq)
5. 25.5 hr
6. 4.03 days
7. 6.41 mCi (237.2 MBq)
9. 330 min
10. 63%
11. 1.32 hr
12. N/2
13. 168.3 (6.23 GBq)

Chapter 4

3. (a) 1707 $\pm$ 13.8 cpm
   (b) 1647 $\pm$ 14.9 cpm
4. 40,000 counts

5. 3 standard deviations
6. 1111 counts

Chapter 5

8. 665 mCi (24.6 GBq)
9. 8.92 mCi (330 MBq)

Chapter 6

15. (a) 7.32 HVLs
    (b) 8 HVLs
17. 10 HVLs
18. 2.31 cm
19. 2 mm

Chapter 8

8. (a) 25%
   (b) 50%
14. 61.4%

Chapter 9

1. (b) 130 cm/min
6. (c) 1911 counts/cm^2

Chapter 10

6. (e) 0.35

Chapter 13

1. 36,541 rad (365.4 Gy)
2. 350 rem (3.5 Sv)
7. 18,144 μCi·hr
8. 1.06 × 10^{-2} μCi·hr
9. 1.2 rad

Chapter 14

4. (a) 0.5 R/hr
   (b) 6.96 mm Pb
6. 1.77 mm Pb
7. 10%

Index